Constructing Cross-

Constructing Cross-Country Obstacles

BILL THOMSON

J. A. ALLEN & Co. Ltd.
1 Lower Grosvenor Place, London SW1W OEL

© W. W. Thomson 1972
The original text and illustrations of this book were
first published in *Stable Management*

ISBN 0-85131-350-7

No part of this book may be reproduced, stored in a retrieval system, or transmitted, in any form or by any means, electronic, mechanical, photocopying, recording or otherwise, without the prior permission of the publisher. All rights reserved.

First published 1972
Reprinted 1973
Reprinted 1975
Revised edition 1979
First paperback edition 1986

Published in Great Britain in 1986 by
J. A. Allen & Company Limited,
1, Lower Grosvenor Place, Buckingham Palace Road,
London, SW1W 0EL

Book production Bill Ireson

Printed and bound by Butler & Tanner Limited, Frome, Somerset

FOREWORD

by Major D. S. Allhusen

THERE is no one better qualified to write a book on the subject of the construction of cross-country obstacles in Horse Trials than Bill Thomson.

His knowledge of building courses to suit every grade of horse from the novice to those of international standard is considerable and extends over many years.

This book is beautifully illustrated and explains in the simplest language the materials to use, the tools to employ and the methods to adopt in order to produce the attractive well built cross-country fence. There is no doubt that our pre-eminence as a nation in cross-country riding is due in no small part to our excellent novice courses in particular, which give such a confident grounding to our young horses and riders in the early stages of their experience.

This excellent manual will considerably help the more amateur course builders, who are principally concerned with pony club and riding club courses, as well as producing some new ideas for the organizers of official horse trials.

I can strongly recommend this book to all those who are interested in this fascinating subject.

CONTENTS

	Introduction	9
1	General Considerations	11
2	Presentation of Certain Obstacles	21
3	Completing Fences	25
4	Steeplechase Fences	30
5	Using Birch	34
6	Thorn Fences	41
7	Problems with some Natural Fences	44
8	Oxers	48
9	Height and Angulation	52
10	The Take-off	55
11	Revetting	58
12	Coping with the Over-wide Ditch	61
13	Uses for Sleepers	64
14	Siting a Table Fence	68
15	The Sleeper Wall	71
16	The Grass Bank	74
17	Walls	78
18	Variation with Logs and Branches	81
19	Coping with Wire Fences	84
20	Uses for Tractor Tyres and Barrels	88
21	Various Forms of Obstacles, including Water	91
22	Multiple Obstacles and Alternatives	94
23	Knots, Ropes, Bolts and other Equipment	98

INTRODUCTION

During the last ten years or so the ever mounting increase in the cost of materials needed for the construction of cross-country fences has enforced alterations to techniques and, indeed, the whole approach to course making and design. However, in revising this book I have not deleted from the text many of the old methods, for there are still situations where they are of value.

Like many sports, cross-country riding has its inherent dangers and as more and more people take part there must become a greater awareness of the risks involved. Course designers and fence makers must try and reduce these dangers to a minimum, while at the same time leaving the character of the competition intact.

1 General Considerations

THIS booklet is intended chiefly to give detailed descriptions of the ways and means of constructing good, solid and durable cross-country fences. As hunter trials and Events become increasingly popular, more and more horses will be jumping the obstacles on each individual course. Therefore it becomes increasingly important that fences should be soundly made in order to withstand a large number of horses jumping over them and to give these horses every encouragement to do so.

A number of factors govern the planning and siting of fences. It may be that only one or two individual fences are necessary. On the other hand, it may be required to link these together to make a continuous course. If it is an individual fence it is a simple matter to choose the site in relation to existent facilities and construct it there. A course requires a deal more thought, and probably the best way to set about designing one is to examine thoroughly the land at your disposal, pick out obviously interesting places—such as quarries, banks, water and the like—then, with the aid of a map, link them together with other more ordinary fences to form a continuous and flowing line.

The First Thoughts
The number of fences constructed in any course really depends on the whim of the owner, but I will mention, for guidance, that the British Horse Society rule that, in Novice Horse Trials, there should be 16 to 20 obstacles on courses varying from 1 to $1\frac{3}{4}$ miles long.

When actually siting a fence the first thoughts must be of the going, and not only of its ability to resist poaching in wet weather but also of its durability. For instance, peaty leaf mould in a wood will quite quickly disintegrate in front of a fence jumped by a number of horses. Horses are always inclined to jump better over natural-looking fences that are blended into their surroundings. If, instead of giving an obstacle wings, it can be sited between two trees or even two small bushes, it will be altogether more pleasing than one stuck up in the middle of a field.

It is impossible even to hint at any sort of specification for fences on difficult terrain, because it is the very peculiarities of the ground that dictate the form which such fence will assume. However, my advice will always be, "Do not try to be too clever!" When designing fences there are a number of points that should be constantly borne in mind. If the fence is solid (does not knock down) it should be so constructed that, if a horse makes a bad mistake while jumping it, he would cause himself the minimum of injury.

Flimsy Rails Encourage Careless Jumping

A course constructor quickly learns never to be dogmatic, but I think it can safely be said that horses do not like jumping out of bright sunlight into deep shadow—a trait which can profitably be utilised by an experienced designer. Drop fences are much more comfortable to jump for all concerned if the landing is still on the downward slope rather than after it has flattened out.

Aquatic sports are good fun, but water into which horses have to enter should come late in the course (not always possible) and should only be used if the bottom, be it in pond or river, is 100 per cent sound and the water not too deep. When constructing obstacles to a specification, they should be measured from the point at which the average horse would take off. This is done by putting one end of an eight-foot-long straight-edge on the highest point of the fence, levelling it with a spirit level and then measuring the distance between the other end of the straight-edge and the ground.

FIG. 1. *Three examples of ground lines.*

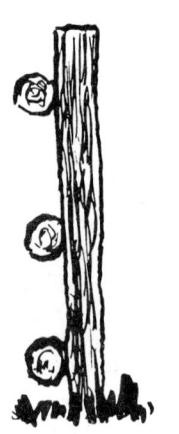

True   *False*   *Helpful*

Before starting to construct or even design fences it is vital to know the experience of horses and riders for whom you are designing them. If they are novices, the obstacles must be relatively simple, inviting and encouraging to jump. For the more experienced they may obviously be bigger, more difficult and pose a greater problem.

For novice horses I believe that all fences should have a ground line, and having said that I now find myself becoming slightly inarticulate in my attempts to define exactly what a ground line is. Fences may or may not have a ground line and if they do it may be true, false or helpful. The true ground line of an upright fence is the line of ground vertically below the top rail. If this line is indicated by a lower rail the fence is said to have a true ground line. The position of the base of the upright posts of a properly constructed post and rails will always indicate the ground line, but obviously a rail attached to them would define it more sharply.

A false ground line usually occurs in such fences as oxers, where a single rail may be constructed—say 3 ft. high and 3 ft. in front of a laid hedge. The visible ground line is now the base of the hedge and that is 3 ft. beyond the top of the first element of the obstacle and invites inexperienced horses to get too close to it before taking off. A similar situation may be produced by leaning a gate towards the line of the course. A helpful ground line is easy to produce—a small tree trunk in front of a post and rails; straw bales; a gorse apron or a well sloped steeplechase fence all encourage a horse to stand back.

The solid parts of a fence should never be concealed by flimsy material such as birch or gorse. The horse must never be subjected to what the F.E.I. call an unpleasant surprise. To me this is a masterly understatement, for a concealed rail can cause a crashing fall.

More and more nowadays course builders are thinking of the safety precautions that can be incorporated into the construction of rigid and solid fences. Several years ago a horse became straddled over the "coffin" at Badminton and as a result of this incident the F.E.I. produced a new rule, the gist of which is that if a horse gets stuck on a fence and needs help the fence judge must tell the rider to dismount and then dismantle the fence as quickly as possible to release the horse. Obviously, once this has been done, the fence will have to be put together again before the competition can be restarted.

Post and rails form the basis of the majority of cross-country fences so it seems logical to examine the detail of their construction first. *They must be strongly made and—just as important—must look it. Flimsy rails encourage careless jumping.* Larch, scots fir, oak, ash and chesnut are all suitable timbers to use. If barked and creosoted they will obviously last longer. Telegraph poles are excellent if they can be obtained cheaply enough. Silver birch is attractive, but the least durable of all. Fences should be as wide as possible, unless, of course, they are specifically narrow in nature, such as a stile. It is suggested that they should be at least 19 ft. long and have four upright posts at 6 ft. centres and average 7 or 8 in. in diameter.

There are four main ways of constructing post and rails, but now it is strongly recommended that for pure post and rails as such, the rails must be lashed into position, so that fences may be dismantled and reassembled in the minimum time. The other three methods I will describe, for they are useful as an integral part of other more complicated jumps such as birch fences and the like. Firstly lay the top rail on the ground in the exact position in which the fence is to be made. Measure out the position of the upright post against the rail, so that if there is any curve in the rail it will still sit tight against the posts. Dig the posts in, and let it be said now, they should be put in two or three feet depending on the going. Perhaps it is worth mentioning that there are several machines in fairly common use for putting in posts. The earth auger digs holes very quickly but once the posts are in them they have to be firmed by tamping. The other sorts actually bang the posts into the ground so that no firming is required.

FIG. 2

On the face of it, the banger would seem to be the better method, but in fact it much depends on the soil. When it is rocky it is difficult to knock the posts in straight. These machines are tractor powered and generally can be hired without much difficulty.

Cut four more posts a little shorter than the anticipated height of the finished fence. Flatten the ground at the base of the dug-in posts on the take-off side. Now wire the shorter uprights to the dug-in posts with 8-gauge wire. Strain the wire really tight with fencing pliers and staple to both posts with 1½ in. staples. Cut two forked branches about 3 in. in diameter and 40 in. from fork to base. The length of the forked branches should be about 12 in. and should spread to an equal distance. With these prop the rail against the double posts and adjust it until you have it at the right height. With a saw mark the wired-on posts at the height of the lower side of the rail. Either raise or lower the rails and cut off the tops of the wired-on post at the marks. Replace the rail and it should fit snugly in position. One point I mention, when laying out the fence, having got the position for the dug-in uprights, mark the rail so that you know which way it was lying, for you will have to move it in order to dig the holes.

When I mention the word lashing to most people they wince and protest that they have never been able to tie knots. A lashing has been devised that has no knots, as such, in it; it is simple to do and is self-tightening. Use plastic cord about 7 or 8 mm. in diameter. The length wanted for each lashing, of course, depends on the size of the timber to be lashed, but it will probably be about 5 to 6 ft. Cut off the required length. Heat the cut ends with a match which slightly melts the plastic and makes the cords stick to each other so they do not fray. Stand at the far side of the fence. Find the middle of the cord and place it round the top of the post with the two ends going away from you. Take them over the rail, round it, then back underneath it to cross each other at the back of the post (Fig. 3). Holding the right-hand end, pass the left-hand end upwards between the post and the rail and inside the cord already in position (Fig. 4). Repeat and still hold the right-hand end firmly, pull the left-hand end tight and thread it round once more and pull tight (Figs. 5 and 6). Now hold the left-hand end firmly and pass the right-hand end in exactly the same manner on the right-hand side of the post. Repeat and pull tight. Repeat on alternative sides of the post until the lashing is really tight. After a little practise a lashing of this type will take less than a minute to tie. If the fence judge is provided with a small chopper the top rail can be removed in no time at all. Remember to leave enough rail sticking out at either end of the fence to take the lashing and, if it is desired to slope the top of the posts, do so after the lashing has been completed (Figs. 7 and 7a).

I have always used a safety device to enable the lower rail of fences to fly off should a horse become hung up. The commonest way for this to happen is that he fails to

FIG. 3

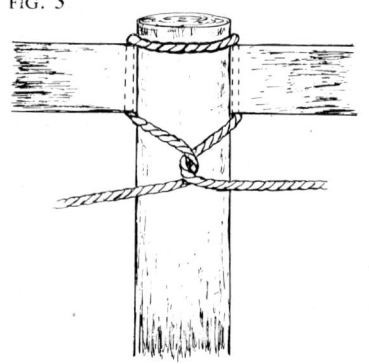

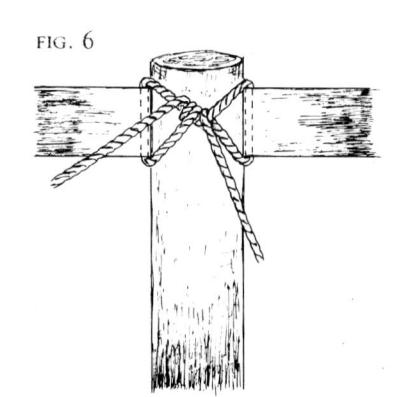

FIG. 6

FIG. 4

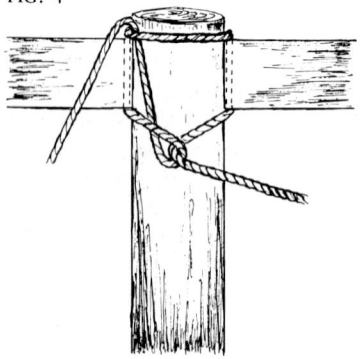

FIG. 5

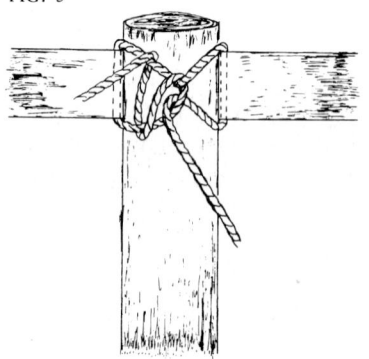

FIG. 7

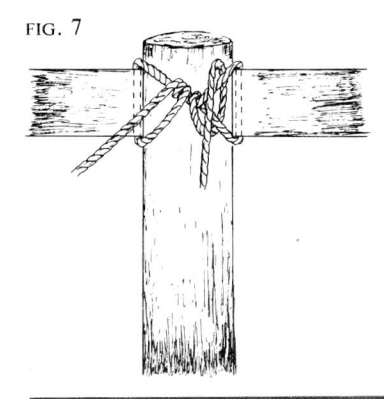

FIG. 7a

clear the top rail with his hind legs and the lower part of them as far as his hocks pass between the top and lower rail. As soon as his hocks strike the top rail his legs straighten out and his fetlocks become jammed against the lower rail. The problem was to make a fence that had a really solid top rail but one where the lower rail would fly off under these circumstances. The lower rail must be of half round sawn timber. From this half round timber cut off two or more pieces about 1 ft. long. The cut is made so that one end of each piece is slanting at about 45 degrees. Prop the lower rail at the right height and nail the pieces of half round into the uprights to support it. If the nails are blunted by hitting the points with a hammer they are much less likely to split the wood than if they were left sharp. This very simple arrangement works well; the lower rail is surprisingly firm unless it is knocked back, then it falls with virtually no resistance at all. I was surprised when the Technical Delegate for the Junior European Championship refused to allow them at Burghley in 1978 and ordered that they should be nailed up tight. I have used them for many years and will continue to do so. (Figs. 8 and 9.)

The lashed fence presents no great problem in construction as long as it is straight, but once corners come into it, the positioning of the posts and props requires a bit more ingenuity. Figs 10 and 11 show the positioning of posts and props on a corner on the near side and a corner on the far side. In Fig. 10 it might be advisable to

FIG. 8   FIG. 9   FIG. 10   FIG. 11

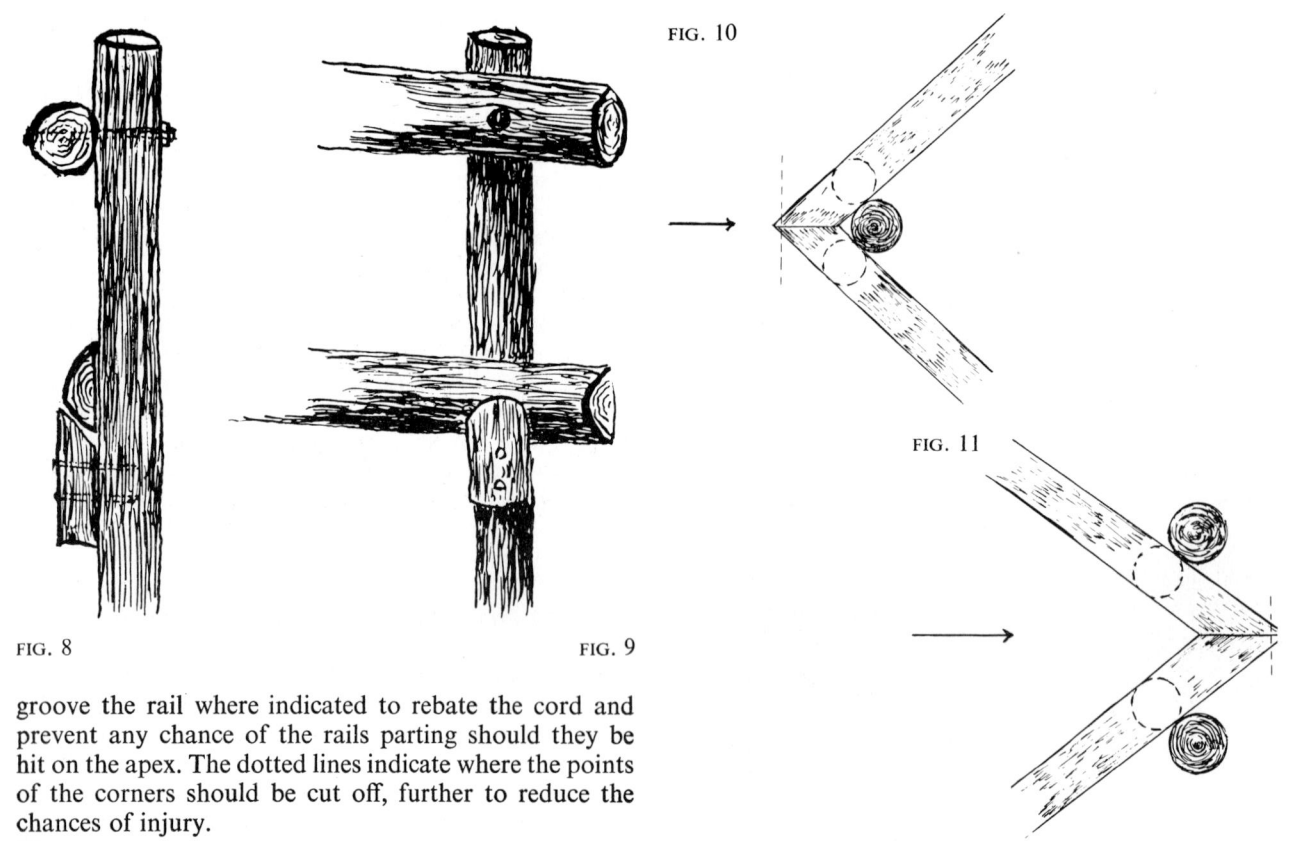

groove the rail where indicated to rebate the cord and prevent any chance of the rails parting should they be hit on the apex. The dotted lines indicate where the points of the corners should be cut off, further to reduce the chances of injury.

Fig. 12 shows another method of putting on a lower rail, this time a round one. It is lashed into position as was the top rail. Lower rails fixed in this way, although they have no immediate safety factor, can be dismantled quickly and they form a much more helpful ground line than the half round rail in Fig. 13. Having returned to the subject of ground lines I would like to explain that they are not necessarily lines in the meaning of the word. Fig. 14 is a hay rack—it was a big one—3 ft. 11 in. high with a 6-ft. spread. These fences without any embellishment have a notoriously false ground line; put straw underneath them and the ground line is immediately

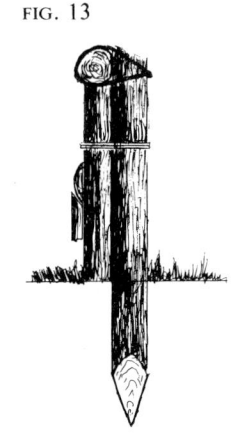

FIG. 12

FIG. 13

FIG. 14

improved. The more straw, the greater the ease with which the fence will be jumped.

For the second method dig the posts in then prop the rail against them and adjust the height very accurately. Now, with a 9/16 in. wagon bit in a carpenter's brace, drill the top rail opposite the posts and put the bit through far enough to mark the posts. Lower the rail and then drill the posts. Use $\frac{1}{2}$ in. coach bolts to fix the rail to the posts.

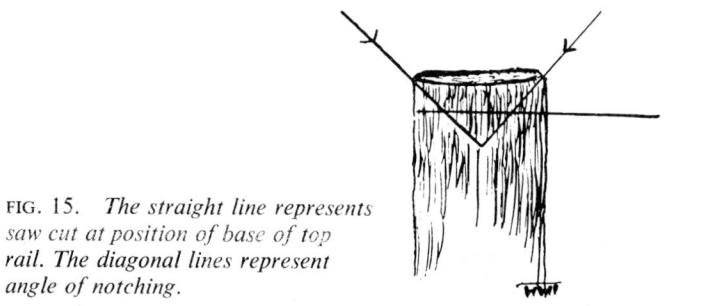

FIG. 15. *The straight line represents saw cut at position of base of top rail. The diagonal lines represent angle of notching.*

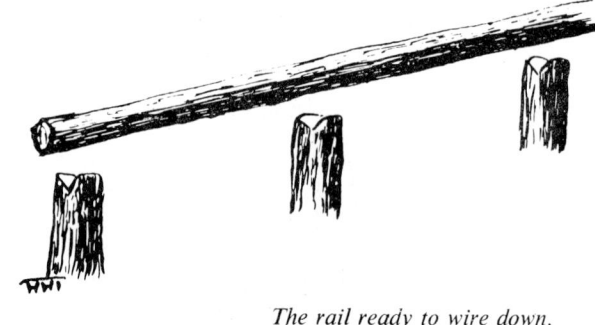

The rail ready to wire down.

Coach bolts may be bought from most good ironmongers from stock, up to 16 in. in length. However, coach bolts are so expensive nowadays that this method will probably not be used unless none other is feasible. Lower rails may be halved, which saves money and may be nailed or lashed into position for they will never receive the punishment that may be meted out to top rails.

For the third method which saves the expense and trouble of obtaining bolts, proceed as for the previous method until the top rail is propped at the required height. With a saw, mark the post at the level of the lower side of the rail, then notch the posts as in Fig. 15. Lay the rail in the notches. Cut some 8 gauge plain wire into lengths and double them. Fix the two ends on the landing side of the posts with $1\frac{1}{2}$ in. staples; pass the loop over the top of the rail to the take off side of the posts. Now put a jemmy or long cold chisel through the loop, drive the sharp end into the post and strain downwards (Fig. 16). This puts a tremendous tension on the wire, which is then stapled up. The lower rail or rails are fixed as before. As we shall see later, this method often forms an integral part of other more complicated fences.

The last and least attractive method is to use halved timber for both rails and posts. The posts are dug in with their flat sides lined up and the flat side of the rail can be applied to them and nailed into position. The result is a fence that looks not too bad from a distance, but is a good deal less convincing from close too.

In methods one, two and four the posts may be sawn off flush with the top rail or left standing an inch or two proud; if they are, the edges should be smoothed with a blacksmith's rasp or a Surform.

FIG. 16. *The method of wiring down.*

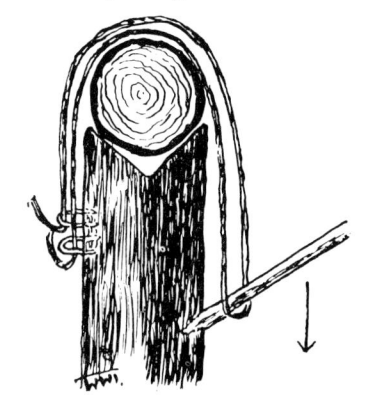

Any surplus length of bolt should be hack-sawed off flush with the nut, and knots in the timber must be removed.

If knock-down fences are required the rail must not knock down too easily or once again carelessness is encouraged. Most people now are conversant with the motor-car tyre gadget for holding top rails, but it is ugly to look at. The steps by which G.P.O. linesmen ascend telegraph poles are not too difficult to come by secondhand. They fix neatly on to posts with coach screws and offer just the right degree of resistance.

2 Presentation of Certain Obstacles

THERE are many ways in which post and rails or double rails may be embellished and their degree of severity increased or ameliorated. First consider the methods by which they may be made more encouraging and simple to jump. Such fences will most probably be included in the first part of a course with the object of getting the horse going. I have already suggested when discussing helpful ground lines, that placing straw bales, a tree trunk or gorse apron in front of the rails achieved this successfully. If the facilities are available for moving a tree trunk, that is fine, for it produces a good jump with very little work.

Straw bales are always available and if they are used they must be fixed; a straw bale flying free amongst a horse's legs can have an unpleasant sequel.

A lower rail should be fixed to the uprights backing the top row of bales, then they and the front row should be transfixed by driving a light stake through them and into the ground. Finally, the stakes are sawn off flush with the top of the bales.

A Birch or Gorse Apron

To produce a birch or gorse apron is a rather more arduous task; they have exactly the same effect as bales but look a great deal more attractive. If using birch, it first must be bundled and tied and should be about 7 to 8 ft. long. The apron extends to about two-thirds of the height of the fence and a rail should be attached to the uprights at that height with a second one halfway between it and the ground. Drive a row of pegs, each about 18 in. long, 9 in. into the ground, 3 ft. in front of the uprights and about 4 ft. 6 in. apart. Cut as many 5 ft. lengths of 10-gauge wire as there are pegs and fasten them to the middle rail, each length being directly opposite a peg and left on the landing side of the fence so as not to interfere with the next operation. The first bundle of birch is laid at the foot of the uprights with the tips protruding some 3 ft. beyond the extremity of the fence. The butts of this bundle are covered by the tips of the next and so on until the whole of the space between the pegs, the foot of the uprights and the middle rail is fitted with birch.

Now pass the wires under the top rail and over the birch and staple each one to its respective peg, pulling them as tight as possible before driving the staple home.

Getting the wires really tight—and they must be—is not all that easy! Having got them as taut as possible by hand, take an end in a pair of pliers and place your foot on the wire and against the stake. Pull hard on the pliers and press down with your foot and 3 or 4 in. of wire will be gained; keep pulling on the pliers as you remove your foot and the bend of the wire against the staple will be sufficient to lock it in position. Repeat the process until no more wire is gained before driving the staple home. To increase the tension further, drive the pegs as far as possible into the ground and saw the tops off them without damaging the wire. Cut the tips and butts of the birch flush with the extremities of the fence with secateurs and trim off any untidy whiskers with a hook. (Fig. 17.)

Another method is to reverse the process and firmly staple the lengths of wire to the pegs; then drive the pegs until they are flush with the ground. Lay the birch as before; then, after positioning the wires, strain them to the higher backing rail with fencing pliers, levering them against the lower edge of the rail. This way is the quicker way but it requires a certain expertise to get the wires really taut.

To make a gorse apron in this manner would require such a tremendous amount of material that it would scarcely be practical; however, a covering of gorse may be given to a birch apron by cutting short lengths of gorse and pushing the stalks into the birch, starting at

FIG. 17

one end and working along, so that the tips of the next piece are covering the stalk of the previous one. This is a tedious business; gorse is not very durable and unless it has been for a big competition, when expense has

FIG. 18

23

been of less object, it has never really seemed to be worth while to use this technique.

A Cheaper and Simpler Way

There is a much cheaper and simpler way of producing a similar appearance, but it will not stand up to the same punishment over a long period, although with a bit of refurbishing it is adequate enough to be jumped several hundred times.

Place straw bales in front of a post and rails. If the rail is about 3 ft. 9 in. high, place them as in Fig. 18, and if 3 ft. 6 in. or less, as in Fig. 19. Fix them as described earlier. Lay gorse lengthwise on the ground in front, of the first row and on top of it in front of the second. Make a series of holes in the ground with a bar about 4 in. apart on the line of the gorse on the ground. Push individual branches into these holes, filling any gaps with smaller pieces stuck into the straw. Cover the top row of bales in a similar way, utilising the string wherever possible. Trim with hedge clippers, dig in a biggish gorse bush at each end to form a wing and the result will be a cheap, quickly made and attractive fence.

Troughs or oil-drums placed at the foot of a post and rails will also produce the helpful ground line, although the latter I do not like. To me they look out of place on a cross-country course, apart from the fact that they are difficult to fix and a loose one would be even more disastrous than a loose bale.

Leaning a plain post and rails away from the line of approach is also helpful in this respect. To ensure that all the uprights are at the same angle, dig the required number of holes in the correct positions and put in and tamp up the end posts only. Then fix the

FIG. 19

top and lower rails to them, so that it is a simple matter to apply the intermediate uprights to the rails and automatically get the correct angle before tamping them up.

3 Completing Fences

HAVING described the ways of constructing birch and gorse aprons, consider how these materials may be used to make complete fences. The straw bale covered with gorse method may be extended to produce an all-gorse fence. The principle of construction is exactly the same. First make a post and rails about 6 in. lower than the ultimate height of the gorse. Build up straw bales until they are slightly above the level of the top rail and then cover them with gorse in the manner previously described. When I have demonstrated this type of fence, people have suggested that it would not stand up to much knocking about. However, I have made a number of them for Horse Trial courses and they have proved to be durable enough; indeed, the first fence at the Everdon Trials has been made in this way for at least six years. There, it is jumped by about 200 horses and so far has not had a whisker out of place at the end of the day. All the same, it is as well to see that there is some spare gorse handy!

The birch apron, too, may be extended in a similar way, but the difficulty to overcome, in this case, is to ensure that the top rail is not only concealed, but *well* covered, too. Construct a post and rails with one lower

FIG. 20

rail fixed slightly higher than half-way from the top rail to the ground, the top rail being about 9 in. lower than the final intended height. If the uprights are 5 in. in diameter or less, bolt a second rail on to the landing side of the posts and about 3 or 4 in. lower than the existing rail. If the posts are of a greater diameter than 5 in., a 1½ in. by 4 in. board may be nailed to the top of the posts; thus, in such cases, a platform is formed upon which the top bundles of birch may be laid (see Fig. 21). The wires are attached to the lower rail, which also supports the thinner part of the fence, passed over the top of the birch and pegged down as for constructing a birch

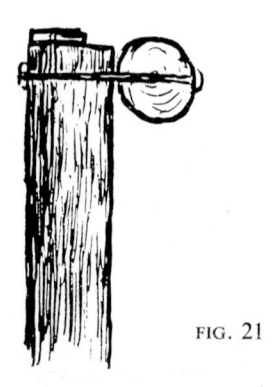

FIG. 21

apron (see Fig. 22). The horizontally laid birch fence is much cheaper and easier to construct than the orthodox steeplechase type of fence with the birch standing upright (see Fig. 23). To make such a fence, say 20 ft. wide, 80 to 90 bundles, of birch will be required, which should be cut not later than March and used the same year. It should be 7 to 8 ft. long with ten or more saplings in each bundle.

The landing side.

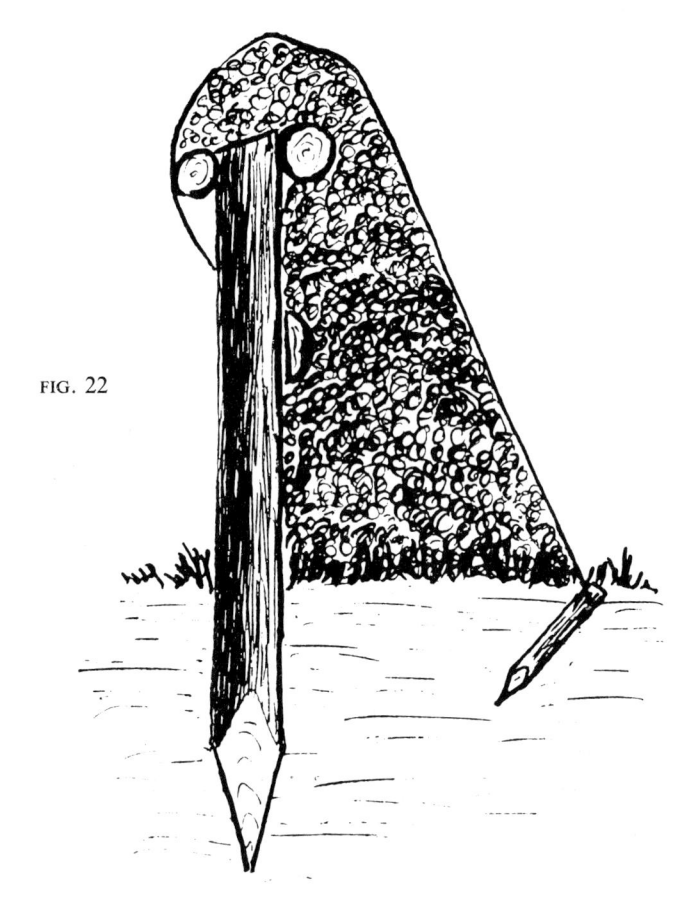

FIG. 22

Keeping the Bundles Upright

If the birch is bought from a forester, it will arrive probably tied once about one-third of the way up, and rather resembling a bunch of flowers in shape. If it is used in this state, difficulty will be experienced in keeping the bundles upright when pulling them into the frame. If you are a perfectionist you will undo the bundles, rearrange the birch so that it all lies straight, and re-tie it, but whether or not they are undone, they ought to be tied at least four times throughout their length so that they more closely resemble a cylinder than a posy.

On the selected site dig a trench, in this case 20 ft. long, 8 in. to 1 ft. deep and 18 in. wide. Drive posts at 4 ft. intervals along both the long sides of the trench, so that they are leaning towards the landing side of the fence. Fix two rails, which may be nailed, to the inside of both rows of posts. Use 6-in. wire nails and blunt them before driving in in order to reduce the risk of splitting the posts. The height at which these rails are fixed depends on the ultimate height of the fence. The F.E.I. stipulate that they may not be higher than 1 metre (3 ft. 3 in.) on a 4 ft. 4 in. fence, which is a good deal higher than normally encountered in this country. Those experienced in making steeplechase fences will probably fix them at 18 in. in front and 2 ft. at the back; for those not so experienced, I would suggest 2 ft. and 2 ft. 6 in. respectively, if the fence is to be not less than 4 ft. Cut the posts flush with the rails and block up one end with two cross-pieces.

Nearest to the blocked-up end of the frame insert the butts of eight or ten bundles of birch and push them up as tightly as you can with your shoulder. Pass a rope from the blocked-up end inside the posts round the birch and back inside the other row of posts, and fasten the two ends to a tractor or Land Rover, and cautiously compress the birch in the frame while an assistant attempts to keep the bundles upright with a crow-bar or sledge hammer. Experience alone will show at what height the rope should be placed against the birch. If it is too low the butts of the bundles will pull in further than the tips and vice versa. Both contingencies must be remedied at once because the fault will be exaggerated as construction proceeds. Having got the first bundles compressed, tie another rope round them and the two uprights of the blocked end of the frame to prevent them from slipping back. Repeat the process until the frame is as full as you can get it. The last two bundles will have to be rammed in by hand before closing the open end of the frame. There is now a formidable barricade of birch which must be clipped to the correct height. The finished fence should slant slightly upwards from take-off to landing and be as level as possible. Use long-handled secateurs, and I would warn you that clipping the birch wears out the sleeves of woollen pullovers faster than any other known method.

The Heights are Level

To get the levels and heights even, drive a post at each corner of the frame. Tie a length of string to one post at the back of the fence at the height you wish, stretch it as tightly as possible and make fast to the second back post. Do not cut it but take it right round the fence and make fast to the original post. In this way the heights are level throughout and the degree of rake may be estimated by looking at the ends. The appearance of the fence is improved if the last 18 in. at both ends are left unclipped. To make it harder and thicker the trimmings may be pushed back and again clipped level.

This may be all that is required for a cross-country birch fence, but if the fence is to be used on the steeplechase course an apron must be added together with a toe board. The cheapest and quickest way to add an apron is first to fix the toe board in position. This is a very strong board, the width of the fence and about 1 ft. by 2 in.; it is pegged firmly into the ground and angled with the rake of the apron. The position of fixing the toe board may be estimated by the fact that a steeplechase fence is as wide at the base as it is high. To economise I now lay straw bales at the foot of the birch fence, but before doing this drive three strong pegs, each with a length of 10 gauge wire stapled to them, well into the ground 18 in. from the base of the fence. Bring the wires back from the fence to get them out of the way while bundles of birch are laid horizontally to cover the straw bales forming a slope from

FIG. 23

the toe board to the top of the fence. Take individual saplings of birch and push the butts behind the toe board so quite a thick vertical screen covers the horizontal bindles. Assistance is now required to have a good strong rail held up against the vertical screen about one-third of the way up the fence. The wires are passed over this, through the fence and strained really hard before being stapled to the back rail of the frame. It is normally sufficient to drive three pegs, one at each extremity and one in the middle, but if the birch is very resistant it may be necessary to use five. Clip the vertical birch saplings to conform with the fence and tidy up the ends of the horizontal bundles at the same width as the fence. At the risk of stressing the obvious, I would draw attention to two points. If either rail of the frame were fixed to the outside of the posts, they—the posts—would make it impossible to compress the birch, and the importance of the trench is that it traps the butts of the birch bundles and prevents them somersaulting out of the low frame when they are hard hit.

4 Steeplechase Fences

In describing the ways in which I myself construct fences, I hope that I do not give the impression that the methods which I use are the only methods, or, indeed, the best ones, for after twenty-three years of fence-making, I still find that there is plenty to learn and new ideas to absorb. Apropos of this it was only within the last twelve months that I learnt that there is a different way of constructing a steeplechase fence from that which I described in my last chapter. I have never seen it used so that my description is second-hand and hearsay only.

To start with, everything is done in much the same way as I have described in previous chapters. A trench is dug to the same dimensions and posts are driven at four-foot intervals along both sides of the trench: but now the difference. Only the landing side rail is fixed to the posts and both ends are blocked up with cross pieces. (In future, instead of using the rather clumsy terms, "Take-off side" and "Landing side", I will refer to the near and far side respectively.) The birch, if in bundles, is then untied and the saplings are put into the trench individually and pressed against the far side rail. This process is continued until the trench is filled. The birch is then compressed by putting pressure on the near-side rail and locking it behind the posts. Obviously, the cross pieces at either end must not be too high on the corner posts or they would prevent the rail from being locked into position. The rail is then nailed up and the fence is finished off in the manner already described. I will not pass comment on which I think to be the more irksome method; both are obviously long-winded and tedious, but I am not convinced that this way of doing it can produce anything like so hard a fence as the other.

While on the subject, it will be as well to consider all types of hedges that may be constructed artificially and then what may be done with natural ones to lower or strengthen them.

An Amusing Obstacle

A Bullfinch is an amusing obstacle, but one that is difficult to simulate. Start by making a post and rails, but instead of using round rails use half-rounds if they are available; apart from being cheaper, they also give a good deal more support to the birch. The top rail is fixed in position at the correct height on the near side and a second rail is fixed on the far side some 4 in. below it. Half-way between the top rail and the ground two more rails are fastened to either side of the posts, both at the same height, and thus a frame is constructed into which the birch may be threaded. The birch should

FIG. 24. *A Bullfinch Fence, showing the method of collecting together the butts of saplings at the base of the fence.*

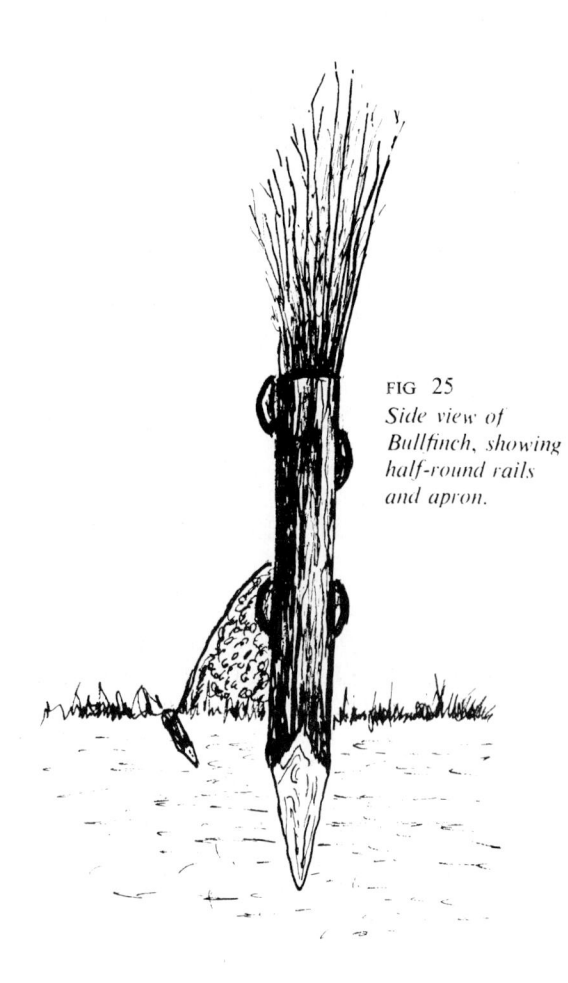

FIG 25
Side view of Bullfinch, showing half-round rails and apron.

FIG. 26. *The "Duck Flighting Butt" at the Burghley Horse Trials, 1962.*

be cut in early March and carefully stored until it is used, which must be in the same year. A screen of birch has now been formed which in no way resembles a natural hedge. If an apron is added to the lower half of the fence the appearance is much improved. Another way of making the birch screen look more natural is to dig a trench between the uprights about a foot deep and as narrow as you can dig it—do this before fixing the rails, of course. Place 4 in. tile drains vertically in the trench at varying intervals of from 1 to 2 ft. apart and refill the trench so that the tops of the drains are flush with the ground level. Now, when threading the birch into the frame, pull the butts of the saplings down into the drains, about six or so butts into each drain. Collecting the individual saplings together at the base of the fence tends to give a better impression of branches growing from a root. The density and height of the part of the fence through which the horse is expected to jump rather depends on the standard of the competition, but the birch must never be so thick that a horse cannot see through it. Spare birch should always be available in order to refurbish in case of damage. An artificial bullfinch will never look like anything else and to get over this difficulty a horse-shoe shaped screen of birch was constructed for the European Championships at Burghley in 1962 and called a duck flighting butt. It was on the lake shore, furbished with spent cartridges, and served its purpose well. If you cannot conceal it, make a feature of it! (Fig. 26.)

FIG. 27

Side view of Gorse Fly Fence constructed with the use of sheep hurdles.

Making a Gorse Fly Fence

The Wylye Horse Trials have an easy and cheap way of making a gorse fly fence employing sheep hurdles. The prerequisite is to be working in a country where gorse abounds. Cut and tie about fourteen biggish bundles, each about 4 ft. 6 in. long and then tie them to two sheep hurdles. Each hurdle will hold six or seven bundles which should be fixed so that they lie alternately with the tips of one to the butts of the next; in this way a uniform thickness is obtained with the gorse standing about a foot higher than the hurdles. Now construct a post and rails with the top rail 3 ft. from the ground and the lower rail 18 in. below it. A hurdle is 6 ft. long, so that the width of the fence should be in units of 6 ft. In this case two hurdles are being used so that the fence would be 12 ft. wide. Place the hurdles against the near side of the rails and tie them there so that they will remain upright. A third rail is now applied to the near side of the hurdles 2 ft. from the ground, and pulled in as tightly as possible and wired to the uprights at both ends and in the middle. The easiest way to get this rail really tight is to hold one end away from the gorse hurdles while the other end is wired up. Then rope the free end to a Land Rover which is cautiously driven back until it has pulled the rail into position, where it is made fast. The wire in the centre may be strained up with fencing pliers.

The top line of the gorse is clipped level. The rake and height of this type of fence may be varied by using longer or shorter gorse bundles and standing the base of the hurdles nearer or further from the base of the rails. (Fig. 27.)

5 Using Birch

I FEAR that this is yet one more of the many uses of birch, and even now I do not promise that I will not mention it again! Despite its versatility, I am not enamoured of it; indeed, I find it the most uncongenial material with which to work. It flicks you in the face, pokes you in the eye, tears your finger-nails and, if wet, soaks you to the skin quick as wink. But despite these unpleasant attributes, I will now describe how it may be used in constructing a prefabricated steeplechase fence.

Prefabricated Fences

To any event which includes a steeplechase phase it must always have seemed an appalling extravagance to make six or so fences, at a very considerable cost, which will be jumped by about 40 horses on one day and then abandoned to the elements for the next twelve months. About ten or eleven years ago, the Badminton Committee realised this and designed a prefabricated fence that could be set up a fortnight before the event, dismantled a week after it, then stored under cover for the rest of the year. These fences resembled very strong show-jumping birch fences and were made in sections 9 ft. long. On the ground two sections were joined together to make an 18 ft. wide fence. At Burghley I copied the Badminton design in which the section will only join together at one end. I have learnt by experience that this is a mistake for, at horse trials, fences of this nature are jumped in the middle which, therefore, takes a great deal more wear than any other part. Obviously, then, if the fences could be joined at either end the duration of their

FIG. 28

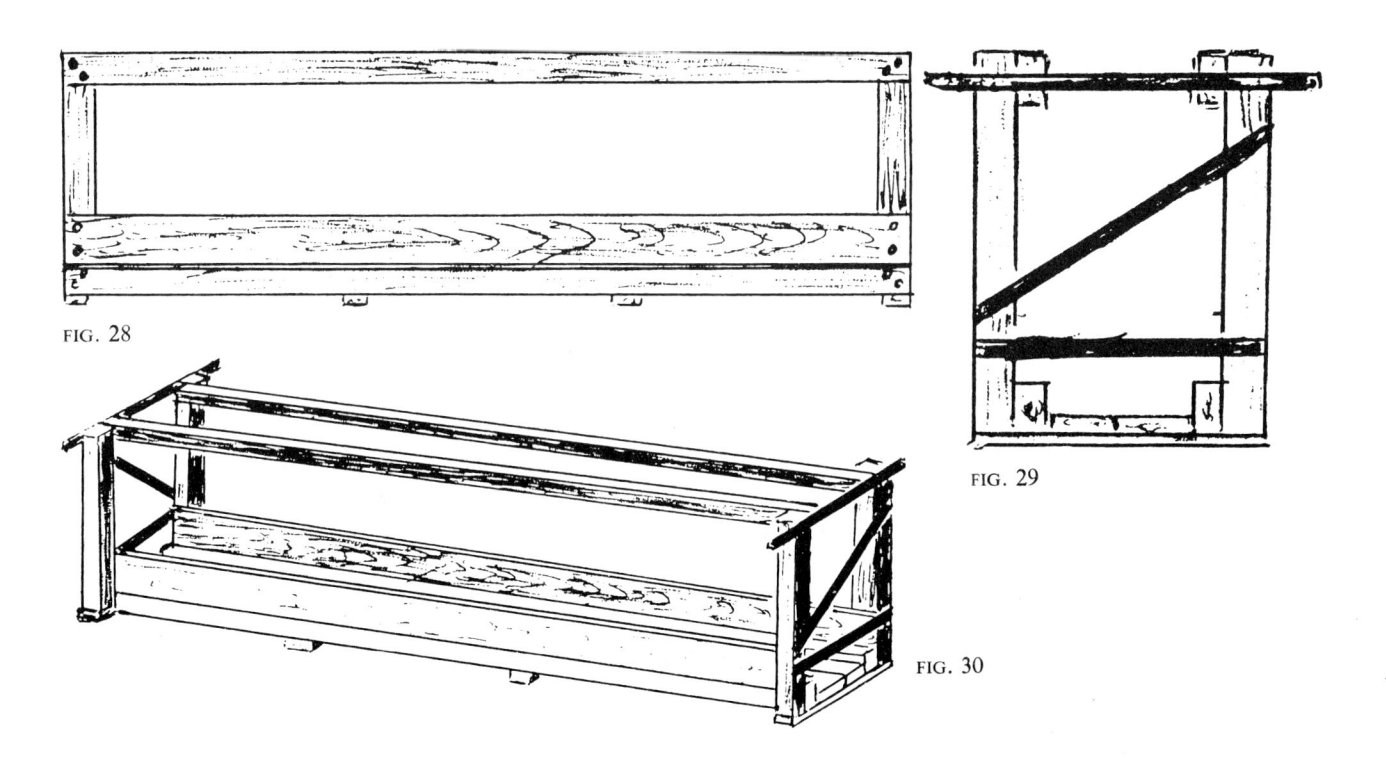

FIG. 29

FIG. 30

life would be doubled. These fences are not intended only to be used on a steeplechase course; they are a very useful stand-by for any cross-country course as well.

Making the Frames
The frames are made from 4 in. by 3 in. sawn timber. First make two rectangles 9 ft. long and 2 ft. 4 in. high,

applying the 3 in. surfaces of the short pieces to the 4 in. surfaces of the long. Immediately above one of the long sides on both the rectangles fix a 1 in. thick board, 9 ft. long by 6 in. wide. All the fixings should be made with $\frac{3}{8}$ in. coach bolts. Fig. 28 shows the inner aspect of the finished rectangles. The frame, excluding the width of the long rails which will be on the inside, will be 18 in. wide.

Iron Bars

From your blacksmith obtain for each frame some 2 in. by $\frac{1}{4}$ in. iron bars: there should be two 32 in. long drilled with $\frac{3}{8}$ in. holes $1\frac{1}{2}$ in. from either end; two 40 in. long drilled with $\frac{3}{8}$ in. holes $1\frac{3}{4}$ in. from either end and two 38 in. long drilled with $\frac{3}{8}$ in. holes $1\frac{1}{2}$ in. and 5 in. from either end. All these holes should be squared in order to accommodate the collar of the bolts which should be inserted so that the nuts are situated on the inside of the frames. These form the bottom, diagonal and top struts respectively. (Fig. 29.) Bolt these struts into position so that the boards are at the bottom of each frame and the upright posts on the outside, taking the utmost care that every angle, save the diagonals of course, is a right angle. Now four pieces of timber 30 in. long by 3 in. wide and 1 in. thick are bolted at 3 ft. intervals across the bottom of the frame, and to them and above them nail the "floor boards", which may be two 9 ft. by 9 in. by 1 in. or three 9 ft. by 6 in. by 1 in. (Fig. 30.)

Stand this completed frame on a level concrete floor and construct as many more frames as you require exactly to match it. Always use the original frame as a pattern and thus any small irregularities will not become exaggerated. It will be seen that the completed frame forms a very similar structure to the orthodox fence frame. The boards at the base form the "trench" which will trap the butts of the birch bundles while the long rails support them.

Filling the Frames

The easiest way to fill the frames with birch is first to fix them in position in the site selected. Place two frames end-to-end and, because the ground will not be level, one of them will have to be blocked up until the holes in the centre iron cross-pieces are exactly aligned. Drive a 3 in. by 2 in. knot-free sawn stake at least 18 in. into the ground at each outside corner of the two frames and one at each side where they join in the middle.

Birch Requirements

The position of these stakes is arrowed in Fig. 31 which is a plan of one frame and part of the second to which it is joined. A $\frac{3}{8}$ in. drill bit is passed through the holes in the iron cross-pieces then through the driven stakes to which the cross-pieces are then bolted. Fig. 32 shows in detail how the cross-pieces of the two individual frames are bolted together.

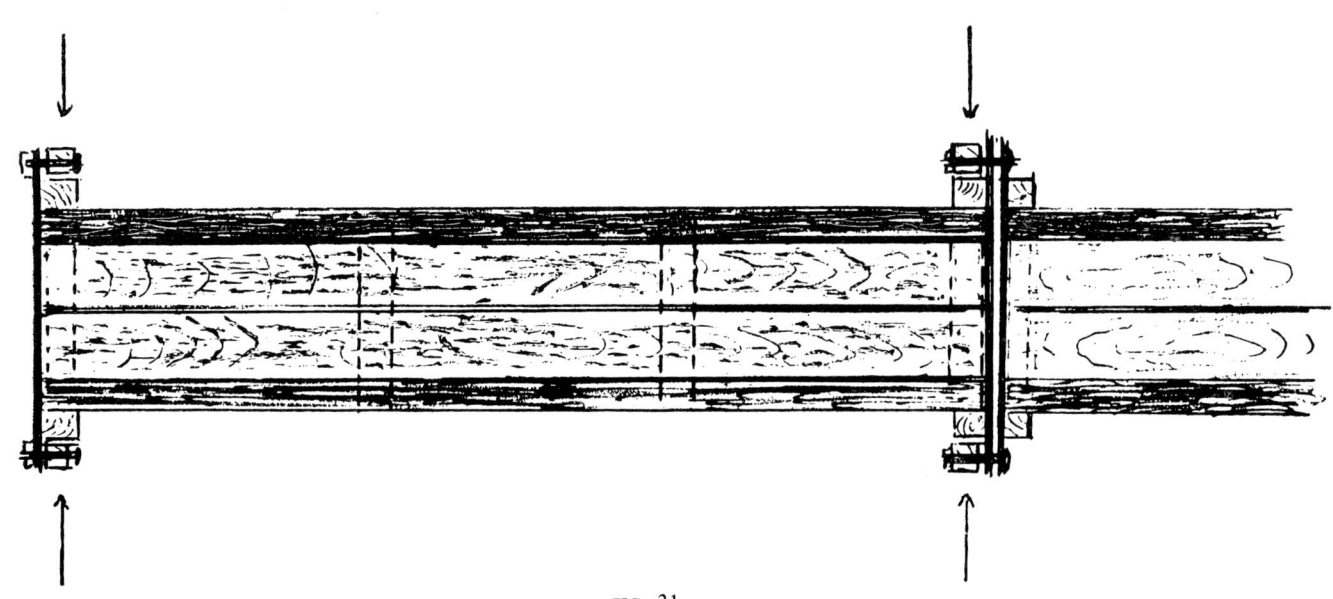

FIG. 31

It is difficult to be dogmatic about the amount of birch required for each frame as the size of bundles varies so much in different parts of the country, but usually about 45 to 50 should suffice.

The birch is packed into the frames in exactly the same way as it is put into an orthodox fence, the only difference being that it is pulled up to the centre of the frames so that the last bundles to go in are at the extremities of the fence. When the first three feet of a frame are packed, wire the two top rails together with 8-gauge plain galvanised wire and repeat at six feet.

The birch is now clipped off with the required rake and at the right height. Again this is done as I have previously described, but great care must be taken to ensure that the line of the top of the fences conforms to the top rail of the frame and not to the ground, for if the fence is on a slope or uneven going it would be easy to clip one end lower than the other, with the result that the frames would no longer be interchangeable.

Wings

The fence is finished off by adding an apron and wings, which may be made by driving two stakes about 4 ft. apart one either side of the fence and nailing four cross-pieces to each pair of stakes. Birch is then threaded through the frame thus formed. My own opinion is that, for Horse Trials anyway, wings should be a prolongation of the fence rather than be angled towards the near side.

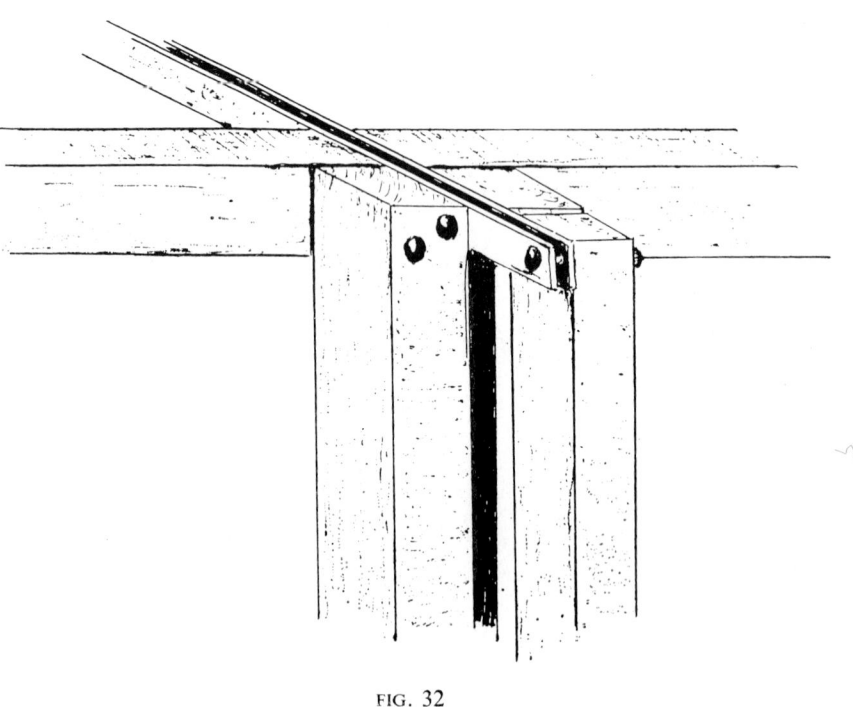

FIG. 32

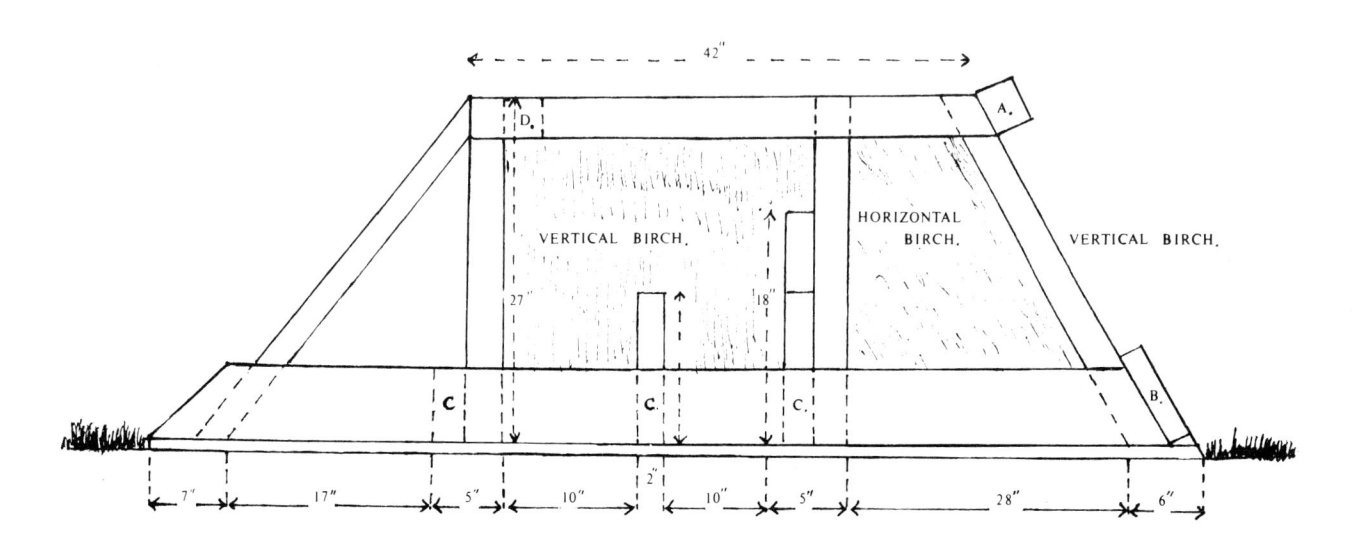

FIG. 33

FIG. 34

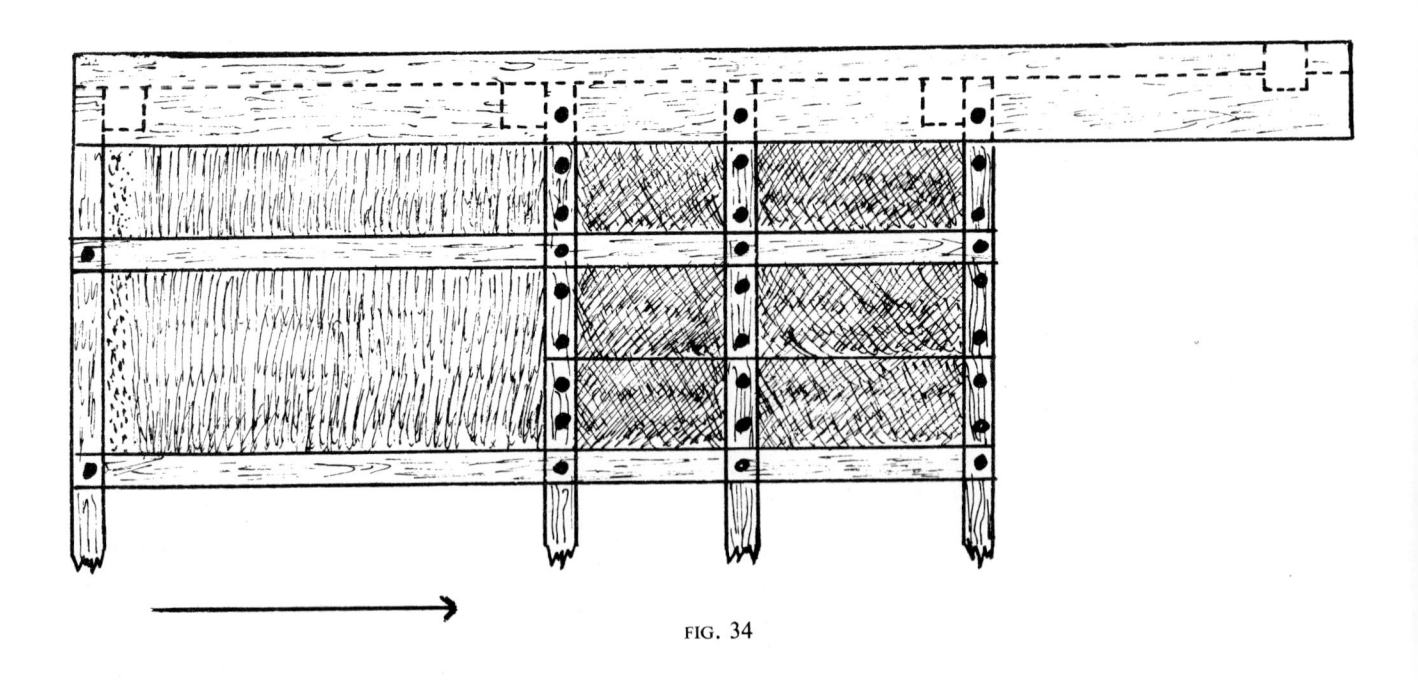

Figures 33 and 34 illustrate a prefabricated steeplechase fence frame that includes a built-in apron. These frames are more difficult to make and when finished they are very heavy. So heavy, in fact, that the sections should not be longer than 6 ft. otherwise there would be considerable risk of damaging them when putting them into position. However, they may be erected very quickly, for apart from bolting the sections together they need no further fixing. The small gaps between the frames must be filled with loose birch which is clipped off to conform with the rest of the fence. Figure 33 shows the end section which is constructed with 3 in. × 3 in. and 6 in. × 2 in. timber. Use bolts wherever possible. The two end sections are joined by "B" the toeboard, "A" the guard rail, "C" the trench boards and "D" the rear guard rail, all of which are 6 ft. long. Figure 34 shows the flooring which is constructed out of 6 in. × 1 in. boards interspersed by 2 in. × 1 in. lathes. The apron is added in much the same way, but it is made entirely of birch. When moving either sort of prefabricated fence, care must be taken to keep it in an upright position for the birch is heavy and might well fall out of the frame.

6 Thorn Fences

THE birch fences that have been described so far make admirable, solid obstacles for both steeplechase and cross-country courses, but they make no pretensions of being anything other than what they are and it is difficult or impossible to blend them into natural surroundings.

In the hunting field, three types of natural, stock-proof hedges are met, and they are often characteristic of the area in which they may be found. The first is the hedge that has been allowed to grow quite naturally but has been clipped back periodically so that it has grown uniformly from the base upwards and forms an impregnable mass of small branches and twigs (Fig. 35). These hedges are quite impossible to simulate artificially. However, hedges are often either cut and laid or staked and bound, and these can be reproduced so that they are indistinguishable from the real thing, but it is a skilled and arduous task to do so.

Naturally, these are generally thorn fences and the basis of forming both of them is very similar. The hedge is allowed to grow until it is 10 or 15 ft. high and then the stem of each individual thorn ("The Grower") is thinned of its more awkwardly positioned branches. After this, about 6 in. or 1 ft. from the ground, it is cut with a chopper to a depth that allows it to split and it is bent almost to the ground without actually

FIG. 35

FIG. 36

being severed from the butt. The operator works away from the direction in which he is laying the fence and, as each grower is bent down, it is intertwined with the previous ones (Fig. 36). A stake and bound fence is further secured by driving split stakes through it and into the ground at 3 to 4 ft. intervals and then long, pliant branches are laced between them to form a continuous plaited line of "binders", about six in number, just above the laid hedge. The skill required in laying a hedge is first in the thinning, then in the depth of cut and finally in the arranging (Fig. 37).

To make a laid fence artificially, first construct a post and rails the required width and a foot lower than the intended height and then screen it with birch. This is done by putting a continuous line of individual saplings into the ground on both sides of the rails. Secure the top ends of them either by stapling them to the rail or by intertwining thin gauge wire through them from one side to the other. Clip them off slightly above the top of the rail. A large quantity of about 10 ft-long thorn or thorn branches must be available, and this is then used to form a screen on either side of the birch screen. The butts must be put at least a foot into the ground, not more than one foot apart and not in a perfectly straight line. The thorn branches are then laid as if they were naturally growing and are wired both to the post and rail and to each other. Push shorter branches into any thin places and secure them likewise. The

FIG. 37

FIG. 38

birch is put in to ensure that the finished fence is absolutely black.

A stake and bound fence is constructed in much the same way, the difference being that the post and rails are constructed without a lower rail and, before the birch screen is added, the stakes are put into position. Use round stakes about 3 in. in diameter and 18 in. longer than the intended height of the finished fence. The last foot of the top end of the stakes should be chopped off on one side only to about half their thickness, which gives the impression that they have been split throughout their length. Drive them 1 ft. into the ground at 3 ft. intervals along the near-side of the post and rails and nail them to the top rail.

Now apply the birch and thorn, laying the latter and securing it. Obtain a quantity of long, straight hazel saplings which should be about 1 in. in diameter at the butt and lace them between the stakes. Go the whole length of the fence with the first, then put in the second, lacing it alternately, and so on until there are five or six in position. With a seven-pound sledge hammer and a great deal of care, tap them down until they are at the correct height and wire them together with light-gauge wire at intervals throughout the length of the fence. Cut off the tops of the stakes some 2 in. above the top binder. Both types of fences are finally tidied up by taking off unruly branches and twigs with a sharp hook.

When completed, these are most effective fences and will last for a long time, apart from the binders, which may have to be renewed from year to year. However, their construction should not be taken on lightly, for even if it is locally abundant, there is more than a day's work to cut sufficient thorn alone.

There are, however, a couple of other ways of disguising a stark birch fence and making it more closely resemble a natural fence. If a fence has been constructed with the birch laid horizontally, it may be faced up with more birch to make it resemble a laid hedge. To do this, about thirty 10-ft. long birch saplings are required and they are put into the ground on the line of the foremost horizontal birch bundles at irregular intervals, never more than 9 in. apart and all angled in the same direction at about 30 degrees to the ground. Their tips are tied back to the topmost bundles which form the summit of the fence proper and then clipped off at that level. It may be necessary to tie them again halfway along their length to prevent them from springing forward. If you are a perfectionist, you will treat the far side of the fence as well. Fig. 38 shows an oxer rail in front of such a fence. If the birch saplings are in leaf, and put in shortly before the competition, a very pleasing effect is produced.

The other trick is to dress the front of the fence by pricking in small branches of thorn until the face of it is pretty well covered. Lock the thorn branches to the birch by wiring on 10 or 12 ft. long butts, cut off, say, 10 ft. high thorn trees, which again should be angled at about 30 degrees. They must be wired on tightly and their tops buried into the birch; to do this means that the wires must be passed right through the fence and pulled to the post and rails at the back. For this operation a long, steel needle is necessary, which any blacksmith will quickly make out of a $\frac{1}{2}$ in. diameter and 3 ft. 6 in. long iron rod, one end of which is pointed and the other drilled with a $\frac{1}{4}$ in. hole to form the eye. This needle is driven through the fence at the required positions with a seven-pound sledge hammer, and when the point protrudes at the far side, light-gauge wire is threaded through the eye and drawn through the fence by extracting the needle from the far side.

7 Problems with Some Natural Fences

HAVING examined the various types of natural stock-proof hedges which are robust enough to be jumped without artificial bolstering and the ways of simulating them, consider the very many non-stock-proof hedges that are met, probably threaded with barbed wire; they present quite a problem on any course.

Figure 39 depicts a truly ragged thorn hedge—if one was generous enough even to call it that! It is difficult to understand how this type of growth is produced. It was a very old hedge which had probably originally been laid then subsequently clipped. The individual trees were too far apart ever to have been stock-proof yet so evenly spaced that there must have been some purpose to their positioning Whatever the cause, the end point is that one was faced with four or five very tough growers, perhaps 4 ft. high with equally tough lateral branches from which sprouted whiskers of small twigs. The particular fence was growing on the top of a fairly steep bank. Its eventual height from base was about 3 ft. 2 in. This meant that all the twiggy bits had to be cut off, which left only some very uncompromising, tortuous stumps with many sharp edges which, if left exposed, could have severely damaged

FIG. 39

FIG. 40

any horse who dragged his legs through them.

One solution, of course, could have been to grub them out, but they were probably well rooted and damage to the bank might have been excessive. It was decided to cut the growers back and surmount them by a rail fixed to posts leaning towards the far side. The latter idea was tempered by expediency, for it is only a glutton for punishment who will dig into the roots at the base of an old thorn hedge. The posts were dug in some 2 ft. out on the near side, to a depth of 2 ft. 6 in., and the holes were wide enough to allow the posts to lean through the hedge. The top rail was then propped against them and, because they were not tamped in, the rail was supported against the growers of the hedge. (Fig. 40.)

FIG. 41a

By adjusting the props the eventual height of the fence was decided. The individual thorns were then cut down to a height just lower than that of the lower side of the rail (Fig. 41a). I am fortunate in using a petrol-driven chain saw which made this operation reasonably easy, but it could be as well achieved with a "Bushman" or any other cross-cut saw. Once the tops of the thorns had gone there was nothing left against which to prop the rail, so it was necessary lightly to tamp in the posts at either end of the fence.

The rail was then propped against them and they were pushed forward until it was exactly over the nearside edge of the cut surfaces of the growers before tamping the posts in really hard. The lower rail—a half-round in this instance—was also fixed to them with 6 in. nails, halfway between the rail and the ground. A neater and more uniform appearance is produced if the butt of the lower rail is fixed, at the opposite end of the fence, to the top rail so long as the fence is constructed on level ground. On uneven ground the reverse may well be the case, but it is only the "eye" of the constructor which can decide. The lower rail was fixed at this stage so that the two centre posts could be held against both it and the top rail to ensure they were in the right position and at the right angle before being tamped in. The lower rail was nailed to them before the top rail was bolted on.

There is a technique which must be followed when drilling straight through both uprights and rails in one go. Before attempting it, ensure that the distance to be drilled is within the scope of the length of the bit, which is probably about 12 in. The bit used in a carpenter's brace is tipped by a threaded point which draws it into the wood, there is then a cutting surface and the chips that it cuts off should be twisted back along the elongated spiral of the length of the bit.

When drilling seasoned timber it works well, but when the medium is new-cut timber, particularly larch or fir, the sap and resin jam the chips into the spiral and the bit winds through without making a clean-cut hole; also, the tension on it becomes so great that it is impossible to drill more than 6 or 8 in. without severe risk of breaking either the bit or the operator's biceps. Under these circumstances, once the bit has penetrated the rail, it will be going forward so quickly that the threaded tip has no chance of gaining a hold on the post and merely pushes it away however hard it has been tamped in. Fixing the lower rail before the upper one helps to support the posts against this contingency, which is never completely eliminated.

In order to be able to drill through upright and rail together, it is necessary to "clear" the hole in the upright, which can be done while drilling is in progress, or after the bit has penetrated the post. The latter method is harder work but less frustrating. Having drilled through the post it is impossible to pull the bit out; it has to be wound out, turning the brace in an anti-clockwise direction. Having removed it, wind it in again for about an inch or so and, while still turning the brace in a clockwise direction, pull it back, drawing the bit out of the hole and with it will come a quantity of the jammed-up chips. Repeat the process inch by inch until the bit will pass freely back and forth through the hole.

The top end of a brace has a rounded wooden knob which fits the operator's hand and which is attached to a metal plate above the top end bearing. When pulling back, pull from the plate and not from the knob, for it is only screwed on and if broken off is difficult to refix.

FIG. 41b

The second method is to clear the hole while actually drilling it. Once the bit has gone in about an inch, pull the brace back and the hole is cleared as drilling progresses. The irritating thing about doing it this way is that the chips do not always clear from the depths of the hole, and when the bit is reinserted the threaded tip is not long enough to penetrate these chips and still get a bite in the solid wood in order to start re-drilling.

When the top rail was finally bolted to the centre uprights, the posts were cut off flush with it and their edges smoothed. Any surplus length of bolt was hacked off. The long grass in front of the fence was hooked away but that at the base of the fence was carefully combed up with a rake to thicken the bottom of the fence and produce a better ground line (Fig. 41b).

8 Oxers

MUCH the same technique is used to protect the thin, twiggy type of non-stock-proof hedge to convert it to an oxer or perhaps a double oxer. There are, however, a number of pitfalls in the construction of such fences that should be appreciated and avoided. In days gone by, an oxer was a fairly common sort of fence in the beef-producing areas of the Midlands. It consisted of a single rail some 3 ft. high, sited the same distance away from the base of the hedge, and its purpose was to prevent the cattle from damaging the hedge. Hence its name. It has already been pointed out that a single rail placed in such a position forms a thoroughly false ground line which should never occur in a Novice course and in a bigger course in exceptional circumstances only. Adding one or two lower rails creates a natural ground line (Fig. 42), or another way of doing this is to add an apron to the base of the hedge so that its near edge finishes directly under the rail (Fig. 43). Adding an apron is perhaps more acceptable on a winter course than a summer one, for if the leaf is out the apron must be greened up just before the course is ridden and there is generally plenty to do at such a time without making extra work.

Simple to Jump

To make an oxer as simple to jump as possible, lean the posts towards the hedge and, of course, fix a lower rail which now forms a helpful ground line (Fig. 44). The height at which the rail is fixed and the distance from the hedge must depend on the standard of the competition, but it also depends to a certain extent on the state of repair of the hedge, its thickness and height. It will be clipped level at

Bill Thomson measuring the cross-country course at Burghley prior to the 1962 European Championships.

FIG. 42

FIG. 43

FIG. 44

whatever height is required and the rail should be at least one foot lower than the top of the hedge. If the hedge is so thin and straggling that it adds nothing to the height of the rail it may be necessary to erect a second rail on the far side forming a double oxer.

In a true double oxer, both rails are fixed at the same height, but I, personally, will not do this unless the hedge is so thin that the far rail can be seen through it and is clearly visible to the horse (Fig. 45). Not infrequently one meets a double oxer on a cross-country course with a perfectly sound hedge between the rails. This, to my way of thinking, is either extremely dangerous or a complete waste of time. If the far rail is

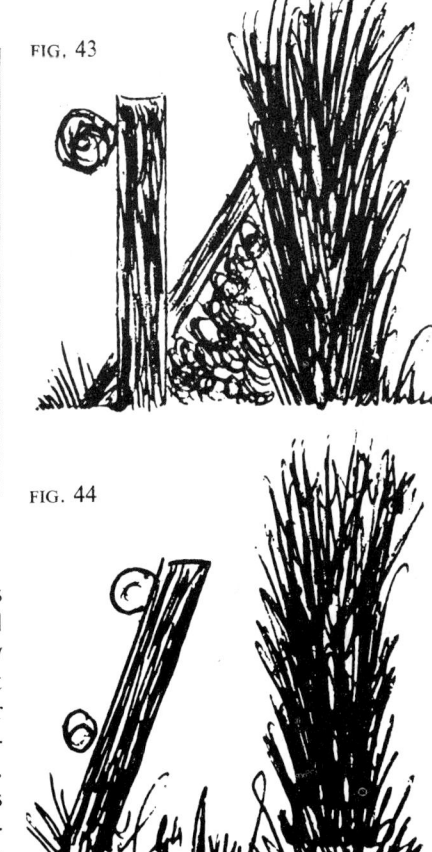

FIG. 45

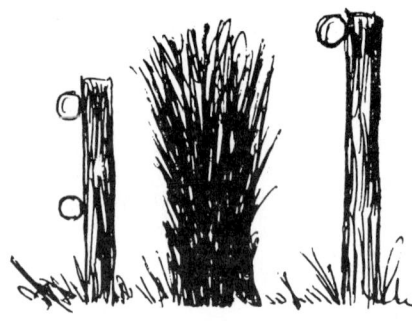

FIG. 46

placed at such a height that it increases the parabola of the leap, then I believe it to be thoroughly dangerous.

A horse that is going well on a cross-country course is jumping his fences but not giving them feet to spare; when he meets such a double oxer he jumps what he can see and his first indication of the presence of the far rail may well be when he hits it. If, however, the far rail is at such a height that it does not increase the parabola of the leap over the visible part of the fence, what is the point of putting it there? Certainly the horse will not give you the answer because it made no difference to him and almost certainly he never saw it! It might be argued that a ditch on the far side of a hedge was no less dangerous nor less of an unpleasant surprise; however, if a horse jumps short at a hedge and ditch he does at least have a second chance to reach for the bank and, if he fails, the result is likely to be far less calamitous than jumping into a concealed rail.

When the Hedge is Too Thin

If the hedge is too thin to be left unprotected but thick enough to occlude vision, it is better to clip it a bit lower and fix the far rail so that it is slightly higher than the top of the hedge which somewhat simulates the show jumping triple bar, and is a pleasant fence to jump. (Figs. 45 and 46.)

To the course constructor it is a matter of considerable aggravation that grassland is frequently populated by cattle, from which he has to protect his fences with posts and barbed wire. It is particularly important to protect birch fences and sometimes it is impossible to finish a fence within the day so it must be wired in. When work is resumed, either the wire has to be taken down or the fence builder has to climb over it every time he needs a tool or another bundle of birch. Figure 47 shows a simple way of overcoming this annoyance. Two staples are driven into each post, one directly above the other. The wire is passed between them and locked into position by dropping a 6-in. nail through the staples. It is a simple matter to pull out the nails and drop the wire to the ground and equally simple to put it back again.

FIG. 47

I make no apology for the great detail in which I am discussing the bolstering up and furbishing of the various types of hedges. When planning a cross-country course I regard all hedges with the deepest affection, for they are natural obstacles in their natural surroundings.

All too often courses have to be made in country which has no natural fence lines, and by natural fence lines I mean any sort of fence other than wire and, rather inconsequentially, that includes walls and post-and-rail fences which should hardly be called natural. On most courses one good, natural place is worth half a dozen island fences. Therefore, when working in open country, any sort of hedge should be seized, worked into the line of the course and rendered indestructible.

51

9 Height and Angulation

THE British Horse Society's rules for combined training stipulate that a birch fence or hedge may be 6 in. higher than the maximum height laid down for any class, so long as this extra height is not solid and a horse may brush through it with impunity. It is easy to achieve this with a natural hedge or the orthodox steeplechase type of fence, if necessary, by thinning out the top with clippers. It is not so easy to achieve in the simulated laid type of fence where the bundles of birch are placed horizontally.

Such a fence, without the embellishments of oxer rails or the like, built for, say, an Intermediate class at 3 ft. 9 in. high, looks a pretty insignificant sort of obstacle and is much improved by adding a 6 in. screen of unresistant birch.

There are, no doubt, many ways of doing this, but I will describe the method that I use, which is fairly simple, durable and produces a good effect. Because the bundles of birch in the main part of the fence are horizontal, the birch in the screen must be acutely angled in the same direction that the bundles are lying. To put vertical birch behind horizontal would produce a very odd effect indeed. According to the length of the fence, construct one or more frames (Fig. 48), using for the ends 2 ft. 6 in. long planks, 1 in. thick and 6 in. wide. The side struts and the diagonal strut should be as long as required and 1 in. thick and 3 in. wide. Nail the horizontal struts to the boards about 3 in. down from the top and the same distance up from the bottom.

One Nail Only

Use only one nail at each end of each strut and place it so that another may be driven alongside later. Make sure at this stage that the frame is a perfect rectangle. Now push the top of the frame one way and the bottom the other so that a parallelogram angled at 45 degrees off the vertical is formed, and nail the diagonal strut in position so that the slant of the frame cannot alter. Then add a second nail to each end of each strut. This frame is wired to the post and rails so that the slant is in the same direction as the run of the bundles of birch.

Pack in individual saplings, starting against the board that is inclined outwards, and keep the birch at that

FIG. 48

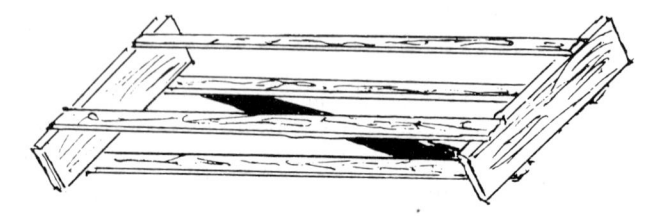

angle throughout the length of the frame. Get in as many saplings as possible, and these should be about 6 ft. long. Do this by hand and do not use any of the methods previously described for packing steeplechase fences, for the birch should not be too hard packed and anyway the frames would not be strong enough to stand that sort of treatment. The top struts should be wired together with 8-gauge wire at 3 ft. intervals and the birch clipped down to the required height. This must seem rather a long-winded way of achieving such a small addition to a fence, but the final result, I think, makes it worthwhile (Figs. 49 and 50).

Two Points to Stress

There are two points to stress. Firstly, it is much easier —for a rough carpenter like myself—to start the frame as a rectangle, and then slant it, than to try and get all the angles right during the initial construction; and, secondly, the intention is that the birch should give as a horse's hooves pass through it, and then spring back into position again. For this reason, the top rails of the frame must not be too high. If they are, the birch may well snap off and a lot of effort will have been wasted!

I think that is the end of the lesson concerning hedges, both natural and artificial, and I would be very pleased to hear from anyone who has new ideas or different methods of construction.

FIG. 49. *This photograph shows the frame and the angle at which the birch is packed into it.*

To my mind, hedges march hand-in-hand with ditches and although in the hunting field ditches are a perfectly natural obstacle, on a cross-country course they generally require a certain amount of bolstering up in order to withstand the depredation of a hundred or so horses jumping them in exactly the same place.

Consider, firstly, the requirements for a ditch on the far side of a fence. It is always possible to increase the width of a natural ditch, but to reduce it on the far side of a fence would be most unwise—indeed, dangerous. This is because it invariably entails revetting, which produces a sheer, hard edge to the far bank which would almost certainly seriously injure a horse that jumped short. Therefore, the first requirement is that such a ditch must be within the prescribed dimensions. The going on the landing side should always come in for serious consideration and, if there is any risk of it becoming poached, it should either be abandoned or given a heavy dressing of cinders at least 6 in. deep and a reserve supply, together with a rake and a shovel, must be close at hand.

Should the ditch be a deep and narrow one, provision must always be made for the means of getting a horse out should he become cast, and a ramp should be dug into it so that, once he has been turned the right way up, it is possible to lead him out.

If a ditch is carrying water, it looks more attractive and, curiously enough, less imposing if the water reaches nearly to the top of the banks, and it is always possible to achieve this by damming it up. However, never do this until the very last moment, for the longer the depth of water is there the more chance there is of it seeping into the ground on the landing side and reducing what was sound going into a morass.

FIG. 50
The finished fence. The white marker is resting on the top of the fixed solid part of the fence. All the birch above it forms the unresistant screen.

10 The Take-off

WHEN a number of horses are jumping in much the same place over a ditch on the near side of a fence it will only be on exceptional going that the take-off does not become considerably cut up and the edge of the ditch will become less and less defined during the course of the competition. Refusing horses, skidding to a halt in front of the fence, will aggravate the situation very quickly.

The Nearside Bank
On a cross-country course it is essential to ensure that the nearside bank of a ditch remains as nearly constant as possible throughout the day. There are two ways in which this may be achieved: either by fixing a take-off rail on the edge of the bank, or by revetting the face of the bank to the depth of the ditch. The former method is by far the cheaper and easier but should only be used where the consistency of the soil is reasonably firm and durable.

For the former method a heavy rail the same length as the width of the fence is required; it should average at least 8 in. in diameter. Telegraph poles are ideal for the job. Apart from being both stout and straight, they have been barked and treated with preservative, which is a consideration, for durability must always be in the course constructor's mind, and a take-off rail which is in contact with the ground throughout its length is likely to rot pretty quickly. Lay the rail in position on the ground, parallel to the edge of the ditch and from 9 in. to 1 ft. away from it. With a spade or turf cutter, mark the ground along the length of both sides of the rail; then roll it back and dig out a shallow trench to accommodate about one-third of the diameter of the rail. Roll the rail back into the trench and fill or remove soil until it lies snugly in position. At each extremity of the rail, drive in two stakes, one on each side, angling them into the bank so that they get the maximum hold (see Fig. 51). If it is possible, the two stakes on the far side of the rail should be driven so that the points are below the level of the bottom of the ditch. Using 8-gauge wire, wire each pair of stakes together, leaving quite a bit of slack, and staple into position. Cut off the tops of the two nearside stakes some 3 or 4 in. above the staples, then drive them so that they are flush with or even slightly recessed into the ground. If, after driving in these stakes, the wire is not absolutely taut, tighten it further by driving the far-side stakes before cutting the tops off them. The cut should be flush with the top of the rail and angled towards the bottom of the ditch. Depending on the width of the fence, repeat

FIG. 51

at about 8-ft. intervals. Remove the turf from the strip of ground between the rail and the edge of the ditch and slope it down towards the ditch. If the rail has been put in position some time before the jump is used it may be necessary to skim it again to remove any new growth of grass. The ditch, too, ought to be cleaned out so that it presents a thoroughly black appearance.

The width of the fence is measured from the near side of the take-off rail to the far bank of the ditch.

Unless a take-off rail is properly positioned and properly fixed, its presence may constitute a greater danger than its absence. If it is situated too close to the ditch or is insufficiently sunk into the ground, a refusing horse sliding forward with his fore-legs braced may break away the ground beneath it and trap his legs under the rail and over the edge of the bank. This situation could also arise if the wires fixing down the rail were insufficiently tightened.

Revetting eliminates the possibility of such a contingency altogether. However, it is expensive and time-consuming but, against that, it will last for a very long time. The principle of revetting is to create a totally rigid wall which supports the near side bank of the ditch so that no amount of refusing or scotching can break down the edge at all.

The best material to use for revettting is undoubtedly railway sleepers, but it can be done with thick planks or even straight poles. The advantages of sleepers are many.

Firstly, they are pretty constant in size. They vary a little in length from 8 ft. 6 in. to 9 ft., but length is the least important dimension. To within a tolerance of $\frac{1}{2}$ in., they are 10 in. wide and 5 in. thick, which makes it possible to build up a neat and tidy wall with the minimum of trouble and of adjustments, and it is possible to decide exactly how many are required to complete any specific fence. They are preserved, and therefore durable. Revetting must be supported on the far side and, whereas it can be done with round stakes, a more professional finish is produced if square-cut posts or half-round timber are used. This makes available a flat, even surface to apply closely to the flat surface of the sleepers.

11 Revetting

REVETTING is not a difficult process but, to make a really neat job of it, a good deal of attention must be spent on the foundation work and, once this has been achieved, the revetting more or less builds itself. It always looks a great deal tidier if it is perfectly horizontal, but if the ditch to be treated runs down the side of a hill, the top of the revetting must conform to the slope of the ground.

FIG. 52

Eight Sleepers

Consider the method of revetting a ditch say 3 ft. deep and, as sleepers will be used, 19 ft. wide. Eight sleepers will be required and five half-round or square-sawn posts. The nearside bank of the ditch is squared off at the required width. If it is too narrow, there will be spare soil which must be disposed of. Most ditches, however, have sloping banks and it may be that the width on the surface is too great while, at the bottom, it is not wide enough. If this is the case, square off the bank at whatever depth gives the correct dimensions and throw the soil back on to the near side. Now, in the bottom of the ditch at the base of the squared-off bank, level the ground to take the first two sleepers and lay them in position.

The next step is to drive in the posts with their flat sides applied to the far side of the sleepers. Position them as in Fig. 52 so that there is a post near each extremity, one in the middle covering the junction of the two sleepers, and the other two half-way along each sleeper. At this stage the greatest care must be taken to ensure that the two sleepers are in a dead straight line and parallel, or as nearly as possible, to the far side of the ditch.

Laying a Long Straight-edge

Having driven the posts to a depth of at least 18 in., the sleepers must then be levelled with each other and, if the fence is on flat going, they also should be horizontal. This is most easily done by laying a long straight-edge—the longer the better but at least 8 ft.—along the top of the sleepers so that its centre is over the junction. It will show immediately if the sleepers are level with each other or if one end or the other needs raising or lowering. A spirit level will show whether they are horizontal or not. Any adjustments must be done at this stage for it is impossible to do them later, and the whole success and appearance of the finished fence depends upon the accuracy with which these first two sleepers are laid. They must be horizontal, level with each other and exactly the same height, so that there is no hint of a "step" from one sleeper to the next.

The Wire should be Tight

Nail the posts to the sleepers and, at the risk of stressing the obvious, that means that the nail goes through the post first. To avoid moving the sleepers, a 14-pound sledge hammer should be held on the near side of them, opposite each post as the nail is driven. If there is not sufficient room to get a sledge hammer behind them, they should be held in position using a bar as a lever. The next four sleepers are laid on top of the original two and the posts nailed to them as before.

At the level of the top of the third layer and opposite each post, dig a trench into the bank about 3 ft. long and 1 ft. wide. At the near end of each trench drive a short stake some 18 in. into the ground and join each stake to the post opposite it with a loop of 8-gauge wire, joining the two ends of the wire by twisting them together. It is important that the wire should be tight and that the ends should be twisted round each other as in Fig. 53, and not as in Fig. 54, where one end is merely wound round the other and

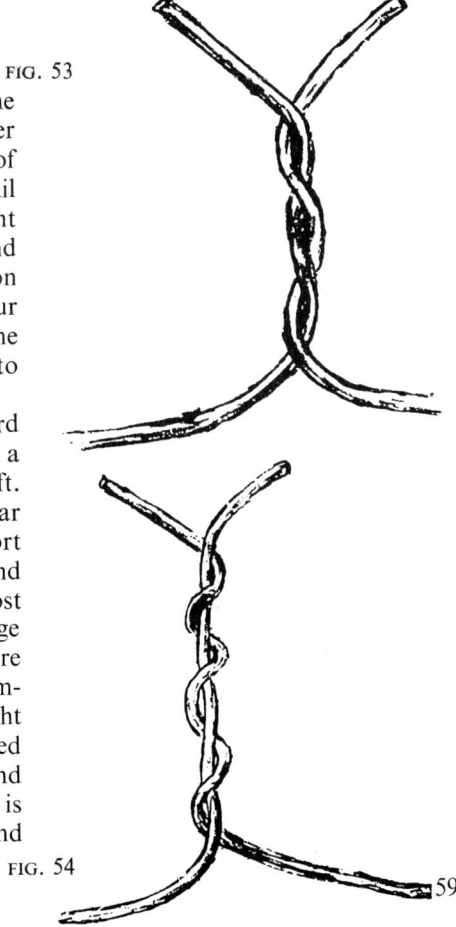

FIG. 53

FIG. 54

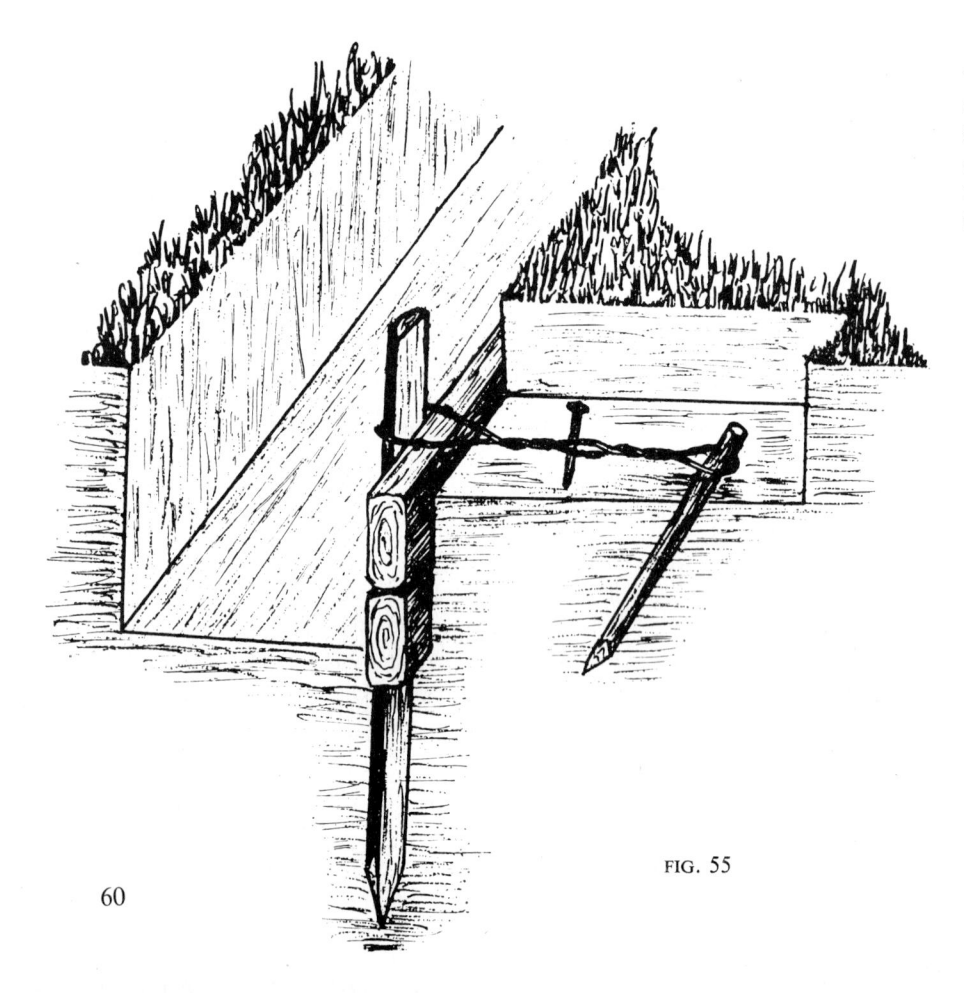

will slip as soon as any tension is applied. With a coach bolt, twist up the wire until either the post or the stake starts to move in the soil (Fig. 55). Make sure that the wire round the posts is touching the top of the sleepers; in fact, it is not a bad idea to staple it there before tightening the wires. Put on the top two sleepers and again nail the posts to them.

A Tedious Business
Now staple the wire to the stakes as near to the ground as possible. Then cut them off some 3 in. above the staple and drive them about the same distance into the ground. Refill the trenches, tamping the soil hard down as the work proceeds, and use the same process for any filling that is necessary behind the sleepers.

No one will make any bones about it; revetting is a tedious business but, if it is done thoroughly and without any skimping, it will last for many years.

FIG. 55

12 Coping with the Over-Wide Ditch

AT the risk of sounding pedantic, I will say that it is always a pleasant task to design and make a course on a bit of country with lots of natural fences, but all too frequently, under these circumstances, one meets with a ditch that is far wider than the allowable dimensions and yet has to be crossed in order to make the best possible course. There are two ways of coping with the over-wide ditch: one is to make it narrower, the other is to bridge it. If it is decided to build a bridge over which a horse may gallop, it must obviously be strong enough to support his weight, wide enough not to frighten him and have a sufficiently good surface never to become slippery.

Obtain three 18 ft. long tree trunks, averaging about 8 in. in diameter, and three 3 in. by 1 in. sawn rails of the same length. Knock off the knots on one side of each of the three tree trunks and to these smoothed surfaces nail the sawn rails. Lay one tree trunk across the ditch with the sawn rail uppermost and dig it in until it is flush with the ground. Do the same thing with the second, leaving 8 ft. between them, and then lay the third half-way between the first two. Align all three of them with a straight-edge and spirit level so that the sawn rails are horizontal and level with each other. This will most probably entail a certain amount of adjustment to the levels of the tree trunks.

Midway from bank to bank, strong stakes—at least 5 in. in diameter—must be driven into the bottom of the ditch. They should be applied to each tree trunk and be bolted to them with $\frac{1}{2}$ in. coach bolts to eliminate any "spring" in the middle of the bridge. Now lay railway sleepers across the three rails previously nailed to the tree trunks and fix them there with 6 in. nails. A solid, safe and firm bridge has now been constructed, but if wet it could be desperately slippery. Therefore it must be covered with gravel, peat or sand, or a combination of any of them. To prevent whatever medium is used from spreading over the sides, some sort of edging must be fixed to the ends of the sleepers, which may be of some 3 in. by 2 in. rough-sawn timber or three more sleepers, one of which will be cut in half. If sand is used as a covering, and it is probably the cheapest choice, it is advisable to nail plastic artificial manure bags over the sleepers to prevent the sand from trickling through the gaps.

Whatever medium is used it should be at least 3 in. deep and must be damped both to increase its adhesiveness and to prevent it from slipping on the plastic bags; or, if sand is used, it will also stop it from being blown off on a wild, dry day.

You can put hand rails on the sides of a bridge. This is done by driving in stakes on both sides of each end of the bridge and nailing another to the sleepers on each side halfway across. On to these, light sawn rails may be attached.

To narrow a 12 ft. ditch to 7 ft., the procedure for revetting is closely followed. At the right position and width lay the first two sleepers, and I cannot stress too often that these are the ones that must be laid correctly. (See Chapter 12.) The wall of sleepers that is being built in this case is totally divorced from the bank of the ditch and thus must be able to stand squarely by itself. To achieve this, posts must be driven in on both sides of the sleepers and bolted together so that the nut will be on the side that is later filled.

It is quite extraordinary what solidarity is achieved when the nuts are tightened but, nevertheless, it is still vital to wire back into the bank, however far away, for there will be a great weight of filling to be tipped in behind the sleepers. Figure 56 shows the first stage of the construction of such a fence, which was at the Crookham Horse Trials. Eight cubic yards of gravel were required to fill the gap between the revetting and the natural slope of the bank. It was possible there to reverse the lorry to the edge of the bank and tip the gravel out.

But such a load of gravel falling on the wires might well have stretched them and nullified their purpose, so, during the tipping process, they were protected by planks laid from the sleepers to the bank. As the gravel was tipped it was levelled out beneath the wires with a spade until it attained their height, and, thereafter, there

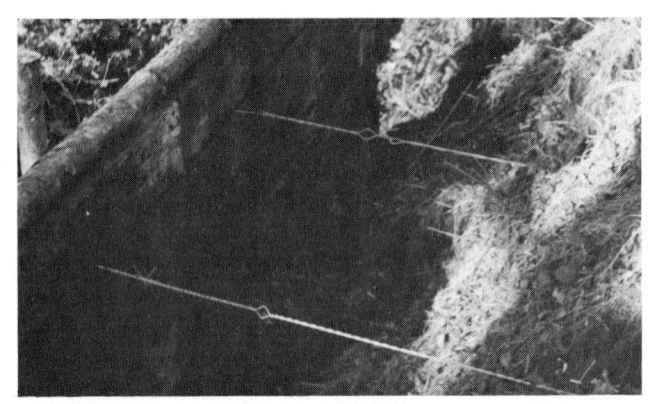

FIG. 56

were no further worries. Figure 56 demonstrates the method of revetting a bank from which a good deal of undergrowth had been winched away, leaving a loose and broken surface. In this particular fence, a ditch had been dug and revetted on the near side of the fence which had been done for two reasons. Firstly, to gain quite a lot of soil for filling in behind the bank, and secondly, to produce the illusion of greater height.

From Fig. 57 it will be seen that the ditch is revetted to the depth of three sleepers which is about 2 ft. 6 in. In Fig. 58 seven sleepers may be counted in the revetting on the bank, which is getting on for 6 ft. Thus the actual leap from take-off to landing is only 3 ft. 6 in. or thereabouts, but when looking at this fence, on foot, one's eye is carried from the bottom of the ditch to the top of the bank, which gives the impression of a very big fence indeed. People may be fooled by these fences, but not horses.

When making a bank such as this or, indeed, when narrowing a ditch, retaining walls must be constructed

FIG. 57

FIG. 58. *Sleeper wall and ditch.*

at both ends of the revetting (as will be seen in Fig. 58) This can be done very simply with two stakes driven in at both ends and rough timber or off-cuts nailed to them. It is wise to line these, too, with plastic bags, which will prevent the filling from falling through the gaps between the boards.

The cost of narrowing this same ditch is not very different from the cost of bridging it.

13 Use for Sleepers

SLEEPERS lend themselves in such a variety of ways for constructing cross-country fences that I have by no means exhausted their potential.

Table fences are attractive obstacles; they produce a spread fence comparable to parallel rails but without the inherent danger of a horse becoming hung up on the second rail. Table fences with a spread of more than about 3 ft. 6 in. should be strong enough to withstand a horse jumping on to them and off again. Strength for cost, sleepers are once again the obvious medium from which to construct such fences. Apart from the fact that they are virtually indestructible, their size and shape lend themselves admirably for the construction of "carpentered obstacles".

A Sleeper Table

Suppose it has been decided to construct a sleeper table 3 ft. 9 in. high, with a 5 ft. spread, and to be made out of sleepers. Such a fence, if it were only one sleeper in width, would look very mean indeed, so it is always advisable to make it at least 18 ft. wide. Therefore, the number of sleepers required are six for the uprights, four for the cross pieces and twelve for the table top—a total of twenty-two sleepers. The site for the fence should be on as level ground as possible, for the finished table-top must be horizontal or else the final product will look completely topsy-turvy.

The first job is to dig in three uprights which, from their centres, should be 7 ft. apart. The two outside uprights should have their narrow sides facing the near side of the fence, whereas the centre upright sleeper should have its broad side facing the near side. These surfaces should be very carefully aligned and the sides of these sleepers must conform to the right angles of the rectangle formed by the table top. The easiest method of doing this (or perhaps, should I say, the least difficult) is lightly to tamp in the two corner uprights and then stretch a piece of string on the near side from one to the other. Before tamping in hard, ensure that the near side surfaces are applied exactly to the line of the string. When tight in the ground, they must also be exactly horizontal and the only certain way of doing this is by using a long spirit level. With the corner uprights now nicely placed it is a comparatively simple matter to get the centre sleeper lined up with them.

The three far-side uprights are put in so that their outward facing surfaces are 5 ft. from the near side surface of the front and their centres are 7 ft. away from each other. It is also necessary to make sure that the corner uprights form a rectangle. Figure 59 shows a

FIG. 59

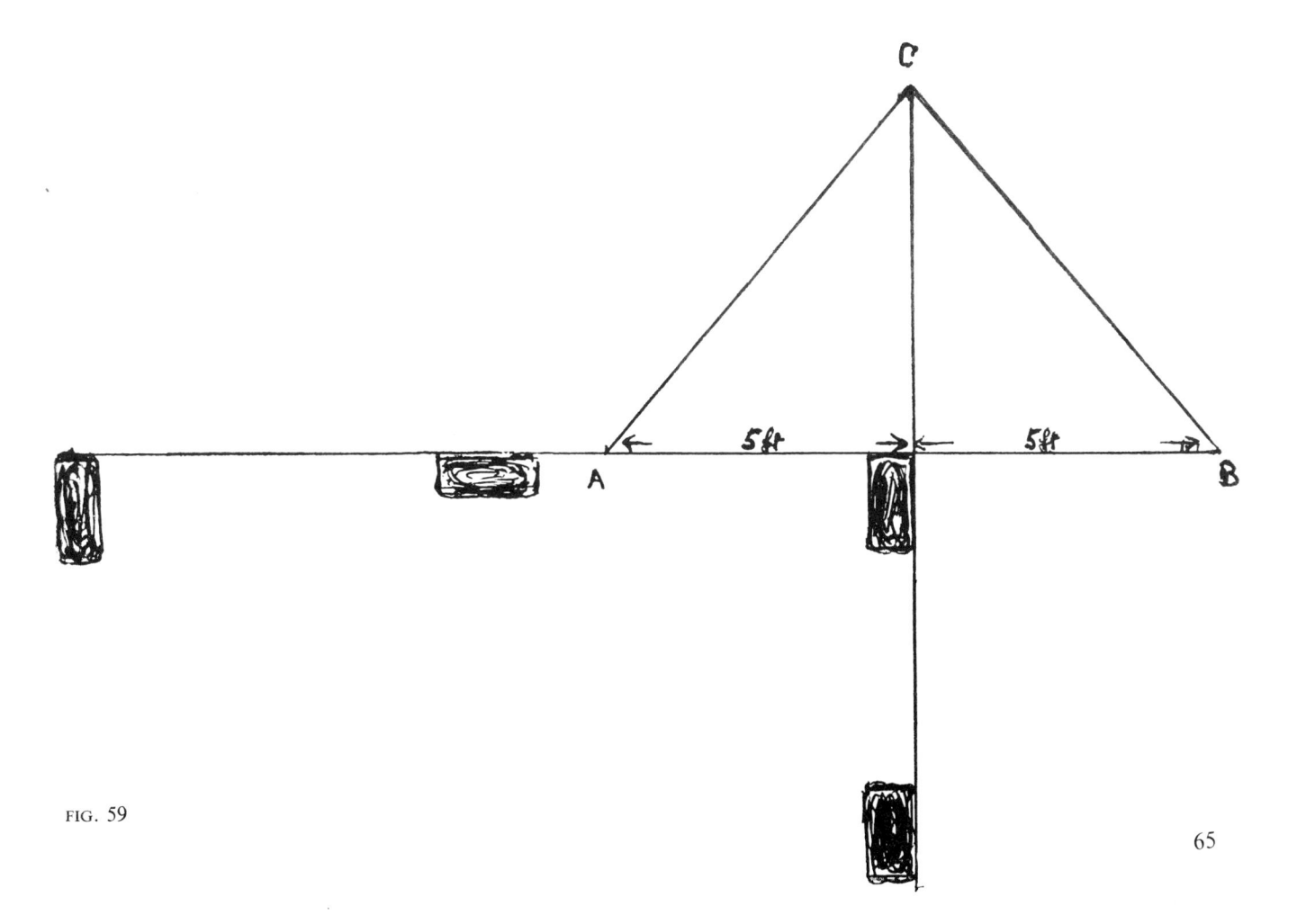

simple way of making sure that a right angle is accurate, as this is difficult to assess by eye. Extend a piece of string along the near side of the fence to 5 ft. beyond the corner upright to "A". Then measure back 5 ft. to "B": drive a small peg into the ground at each of these points. Cut another piece of string about 15 ft. long—its exact length is immaterial. Tie a small loop at either end, put both loops over one of the pegs and, with the string thus doubled, find its centre. Having found its centre, put one of the loops over the other peg and walk back until both halves of the string are tight, which will put you at "C". A line from "C" to the corner sleeper will cut the original piece of string at right-angles. In this way, the far-side corner uprights may be accurately positioned.

Cross-pieces will be fixed to these uprights and the sleepers that will eventually form the table top will be laid on the top of the cross-pieces. Thus the uprights must be cut off 5 in. below the ultimate intended height of the fence, that is, so long as the site on which the fence is being constructed is horizontal. If, however, the ground is sloping but the table top is horizontal, it is obvious that, over a length of 18 feet, one end will be higher from the ground than the other.

Optimum Height at the Centre

It must, therefore, be decided what the optimum height at the centre is going to be. Mark the near side centre

FIG. 60

post at the intended final height, which is probably best done by holding a rail against the uprights so that the variation in height at the extremities may be gauged. Re-mark the centre upright 5 in. below the original mark. Then, using a straight-edge and spirit level, find the position at which the near-side corner uprights must be cut. Cut all three near-side posts, taking the greatest of care that the cut is level and horizontal. Again using a spirit level and straight-edge, from the near side uprights to the corresponding far-side uprights, find the

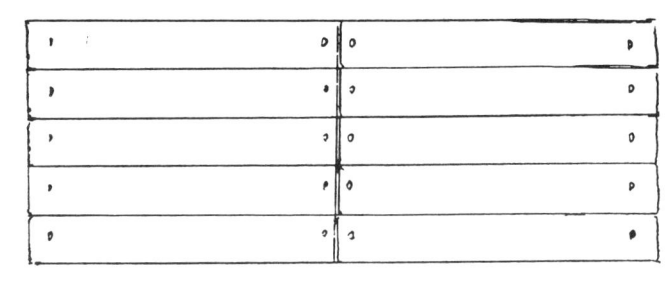

FIG. 61

correct height at which to cut them and then saw them off.

If this work has been done accurately, the fence may be completed with little further trouble. If, however, the uprights do not form a rectangle at the right dimensions the finished fence will look—at best—untidy.

Bolt two cross-pieces to the top of the corner uprights on their outer faces, using two $\frac{1}{2}$-in. coach bolts to each upright. They will probably have to be 11 in. long. I say "probably", for sometimes, where the railway ties have been bolted to the sleepers, the weight of the trains going over them has compressed the wood by as much as $\frac{1}{2}$ in. The centre uprights should be shouldered as in Fig. 60, so that they provide a $2\frac{1}{2}$ in. bearing surface for the centre cross-pieces. Drill right through both cross-pieces and the upright and bolt together with $\frac{1}{2}$ in. by 16 in. coach bolts.

An Arduous Task

Cut the cross-pieces flush with the uprights and lay the ten top sleepers on them as in Fig. 61. Drill each sleeper above each cross-piece, then fix them in position by driving an unsharpened $\frac{1}{2}$ in. by 8 in. coach bolt with a seven-pound sledge hammer. Not infrequently, sleepers have a naturally bevelled edge and it is well worth while, before constructing a table fence, to look over the sleepers and pick out four with such a bevel and save them for the near and far side edges of the top.

A word of warning! Cutting sleepers by hand is an arduous task; it is, of course, much easier with a chain saw, but sleepers often have small stones embedded in them and the damage these stones may do to a chain is quite remarkable.

FIG. 62

14　Siting a Table Fence

Figure 62 depicts a table fence made from sleepers and sited on sloping ground. It was made for the Farleigh Castle Horse Trials, which was a novice event. In such events the height limit for solid fences is 3 ft. 6 in. The right-hand end of this fence—perhaps it would be wiser to say the lower end, for those who are observant enough will have immediately noticed that horses will be jumping it towards the camera—was about 2 ft. 6 in. high,

while the higher end was over 4 ft. high. The optimum height was attained about one-third of the way between the centre and the higher end.

Impression of Width

Without any embellishment, this fence was a sheer waste of time for, if it could be jumped at the lower end, it produced a laughably small obstacle. On the day, the lower end was blocked out by standing milk churns on it. The result was that, firstly, the fence was called a churn stand and looked reasonably naturally sited. Secondly, it was impossible to jump it at the low end and nobody in their right mind would take on the higher; so that the obstacle became a pretty constant 3 ft. 5 in. in height. Thirdly, although the width of the fence that was jumpable was fairly narrow, the impression was of a wide obstacle. If the lower end had been blocked out by artificially planting a tree or bush of some sort on the near or indeed the far side, the same result would have been achieved but the impression of width would have been totally lost.

Table fences made from sleepers may form the basis of a number of specific obstacles depending on the ingenuity and imagination of the course builder. If some object such as the saw bench in Fig. 63 is going to be mocked up, it must be done well and with as much realism as possible—a badly finished fence of this sort stands

Direction of Rotation

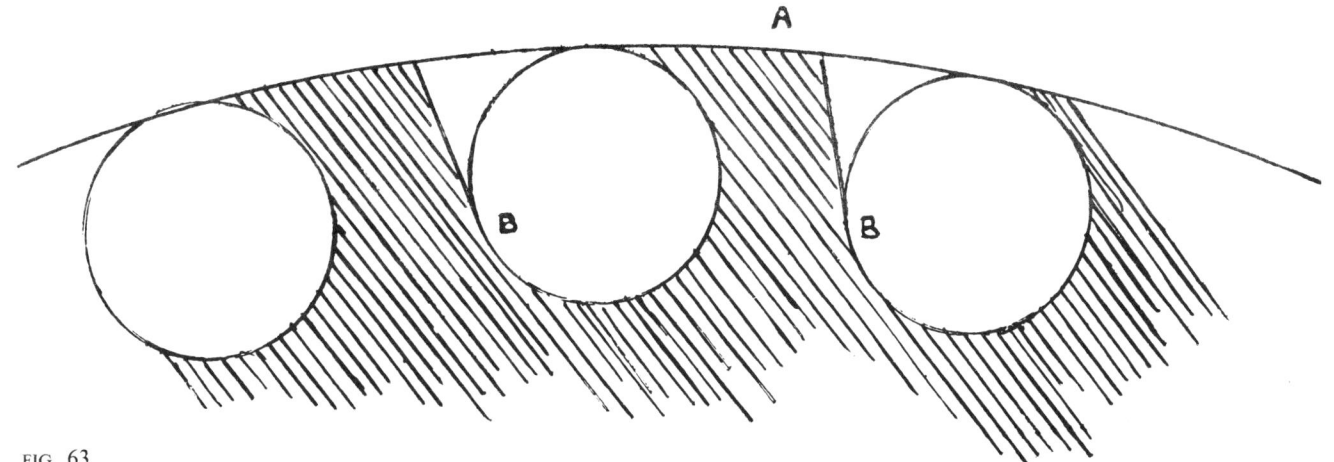

FIG. 63

out like a sore thumb. Constructing this saw bench, which was used in the European Championships at Burghley in 1963, was great fun and it provoked quite a lot of comment. Indeed, the saw blade was so realistic that spectators could not resist the temptation to see if the teeth would bend—as it was made of hardboard many of them did!

Truly Horizontal

The construction of this fence exactly followed the instructions given for a table fence made on sloping ground, the only difference being that it was below the optimum height—in this instance 3 ft. 11 in. Even at the higher end it was subsequently raised to 3 ft. 11 in. throughout its entire length by fixing a tapering length of timber to the top, which served two very useful purposes.

As I mentioned, the top of table fences must always be truly horizontal, regardless of how the ground may be sloping beneath it.

69

By fixing this tapering piece of timber to the fence both a horizontal top and maximum height were achieved. Secondly, this piece of timber supported the saw blade. This was done by cutting a slot in it to the required length and dropping the blade, which was only a half-circle, into it. It is essential in a competition, especially of the importance of a European Championship, that fences must remain absolutely constant throughout the day. Had a horse hit the saw blade it would have broken, so it was therefore necessary to have a spare blade and a method of quickly replacing a broken one.

Not too Arduous

To disguise the sleepers, the table top was covered with hardboard, which was held down at the edges by bolting 3 in. by 1 in. planks through the sleepers. Two boxes were constructed beneath the top to simulate the engine housing and pulley bearings, and the whole structure was painted grey. The pulleys themselves were short logs stripped of their bark and bolted into position, and the belting was made from a bit of old canvas hose-pipe. A liberal quantity of sawdust was heaped in suitable positions to add to the realism.

FIG. 64

The most difficult part of the fence to make was the saw. The method eventually used was not too arduous and produced a remarkably realistic result. First, concentric circles were drawn on a bit of hardboard. The larger one to the dimensions of the finished saw and the other half an inch smaller. This was then marked off with a compass every two inches. At each of these marks a 1-in. hole was drilled with a high-speed wood bit in an electric drill. The circle was then cut in half and on both halves the line of the larger circle was cut round with a fret saw so that each semi-circle had an edge patterned as Fig. 63. Then, with a tenon saw or a fret saw, each tooth was cut down the line AB and the whole thing was painted with aluminium paint.

The fence depicted in Fig. 64 was, in fact, a double, indeed the numbers

26 and 27 on the white flags are just discernible. 26 was the saw and 27 was a tree trunk which had been cut into slabs to season and was borrowed from the saw mills. There are advantages for using a tree trunk in this form but firstly I wish to stress the desirability of having everything in keeping, so what more natural than a newly-sawn tree trunk lying beside a saw bench?

15   The Sleeper Wall

I HAVE not specifically described how to make a sleeper wall, although the method for narrowing a ditch follows the same procedure. A sleeper wall only 9 ft. wide—the width of one sleeper—looks a very mean little fence and sleeper walls should be made not less than 18 ft. wide.

Lay two sleepers end-to-end on the ground on the desired site and level them with a spirit level, which may entail letting one or other of them into the ground. Drive in strong half-round stakes, with their flat sides applied to the sleepers, at each extremity half-way along each sleeper and where they join in the middle. Now build up the sleepers to the required height, which may necessitate the bottom sleepers having to be dug into the ground. Then drive in five more posts opposite the ones already driven to trap the sleepers between them. Drill the posts through the top sleepers and bolt them together, and nail through them into the lower sleepers to prevent them from moving sideways. This method produces a neat and very rigid structure. (Fig. 65.)

FIG. 65. *Sleeper wall.*

A Different Sort of Obstacle

Artificial banks are both time-consuming and expensive to make and, even after they are made, their maintenance needs quite a lot of attention. But they are a very different sort of obstacle. Banks fall into two categories: those that may be "flown" and those that have to be negotiated by jumping on to them and off again, but never should the course designer's intentions be in doubt. Either they must not exceed the maximum spread laid down for any particular competition, or they should be so wide that it is impossible for a horse to leap them in one. In the former case it is always possible that a horse may change legs on the top so that such a bank, of whatever material it is made, should be strong enough or sufficiently well consolidated to withstand a horse's weight. Banks of both categories may be made out of turfed soil, or they may be either stone or sleeper faced.

The sleeper-faced bank is really just a box made of sleepers and filled with rubble and soil and surfaced with turf or cinders. Firstly mark out the outline of the box on the ground at the required dimensions, with pegs and string. Make sure that the outline is a true rectangle by using the method described in Chapter 14. Lay the first tier of sleepers along the string, align them carefully and level them, using a spirit level.

Drive half-round posts, on the inner aspects only, of the near and far side sleepers and on the outer aspects of those on the extremities of the box. Drill the near and far sleepers opposite each post and bolt them so that the nuts are on the inner aspect of both lots of posts. Repeat the process with the second tier.

Consolidate the Material

The walls of the box at this stage will be about 20 in. high and it should now be filled to that height with whatever medium is to be used at this stage; stone or rubble is perfectly satisfactory. Put it in carefully to ensure that the posts are not pushed outwards. Cut five lengths of 10-gauge wire slightly longer than twice the spread of the box and at the level of the top of the second tier of sleepers. Join each upright on the near side of the fence to its opposite number on the far side so that there are five loops of wire spanning the spread of the fence. Tighten these loops by twisting them as described in Chapter 12, and do the same thing to the posts retaining the ends.

Consolidate the material that has been used for filling before bolting the third tier of sleepers in position. These should then be filled and wired as before, again consolidating the filling material as hard as possible. The bank is then finished by putting more sleepers in position until the desired height is obtained. A four-tier bank will be approximately 3 ft. 4 in., whereas a five-tier bank will be about 4 ft. 2 in. high. The easiest way to adjust the height is to dig in the bottom tier of sleepers. If the ultimate height was intended to be 3 ft. 9 in., then obviously the bottom tier of sleepers must be let in to a depth of 5 in.

A Sand Bunker

The material used for the top filling depends on whether the bank is to be turfed or not. If it is to be turfed, there should be at least 1 ft. of well-consolidated soil on top of the rubble and the bank should be turfed in the late autumn, winter or early spring. If, on the other hand, it is decided to use cinders, smaller stones must first be put on top of the rubble and then cinders on top of them. If this is not done, it would be a number of years before the cinders had permeated the rubble and, during that time, the level of the top of the bank would continue to sink. Indeed, it is always wise to leave a bank for some months before surfacing it.

It may be decided to dome the top of a bank or make the far side higher than the near, in which case turf will have to be used. The only way to make a grass surface that will last for any length of time is to cut strips of turf about 6 in. wide and as long as possible and lay them on their sides so that the roots of one tuft are applied to the herbage of the next. This method takes a lot of turf, but it is worth it in the long run.

Figure 66 is a photograph of a sleeper box made to

FIG. 66

resemble a sand bunker filled with gravel. The sleepers were covered with hardboard, the panels were made from picture rail moulding and it was painted to resemble concrete. It had a spread of 6 ft. 6 in. and was 27 ft. wide. It was situated by the drive to Burghley House and, therefore, had to be dismantled after the trials. In fact, it was really a table fence with a 1 ft. high edging fixed to the table so that about one cubic yard of gravel was sufficient to fill it.

73

16 The Grass Bank

To construct an all-grass bank which is only retained by sleepers at the extremities is a pretty big undertaking which will also be expensive and involve a great deal of work. The dimensions of such a bank could, of course, vary according to the type of competition, but I cannot imagine anyone making one where it was not going to be jumped by Advanced class or International horses; in this case it would probably have a spread of some 15 to 20 ft. at the base and rise in a symmetrical mound to some 5 or 6 ft. at the highest part. However, the point on which a horse ought to land and retain his footing should not be more than the maximum height allowed.

Inclusion of a Ditch

Figure 67 shows the outline of the cross-section of a grass bank and the arrow indicates the place at which a horse is likely to touch down. The fence is made much more interesting by the inclusion of a ditch on the near and/or far sides, which adds a bit to the soil required to make the mound, but not very much. In the first instance, two sleeper walls must be constructed at the site of the extremities of the fence. They must be exactly parallel and, indeed, form the ends of a rectangle and they should be at least 20 ft. apart.

Now it is a question of tipping soil between these walls until the height of the mound is such that it is no longer possible to tip. The walls should be wired together at intervals and the sides of the bank shaped with a spade as the work progresses. Every load should be consolidated as hard as possible.

When it is no longer possible to tip it is advisable to dig out the ditches and throw the soil from them on to the top of the mound. These ditches will almost certainly have to be revetted. Any more soil that is required is now added and the mound finally shaped.

The heart of the bank may well be filled with rubble or any other non-compressible material, but it should be covered by at least 3 ft. of soil. To get a satisfactory surface on the steep parts of the bank, at any rate, turf should be laid in strips. The flatter surface on the top may be turfed in the orthodox manner. A fence such as this should last for many years, but it does require a certain amount of attention.

Treated with a Selective Weed-killer

Firstly, it will be at least six months before it can be jumped, by which time the soil will really have begun to consolidate and the turf should have put up a reasonable growth. It should be treated with selective weed-killer

FIG. 67

and lawn sand until the turf is properly established; thereafter, the grass should never be allowed to grow really long and every time any wear appears it should be repaired straight away.

Under certain circumstances, if a bulldozer is available, it is possible to push up a bank with much less work. If this is done on the flat, a great deal of surface soil will be pushed in from all round the site of the fence and this area will have to be cultivated and re-seeded. If, however, a field of deep ridge and furrow can be found, the damage done to the surrounds of the fence is greatly reduced. Site the fence on a ridge—the bigger the better—and, where the retaining sleeper walls are to be made, dig two trenches one spit wide and to the depth of the bottom of the two adjacent furrows. Into each trench lay two sleepers, end-to-end, and make sure, using a straight-edge and spirit level, that they are horizontal. On these first sleepers build up the two retaining walls to the correct height. Remove the turf between the walls and over the same width on the next ridge on both the near and far side. An experienced bulldozer driver should now be able to gradually push up the soil from these two ridges on to the top of the ridge between the retaining walls.

Using the Wire as an Anchorage

It is impossible to shape the sides of the bank while the work progresses, but, to start with, a great degree of consolidation may be achieved by motoring the bulldozer over the soil that has already been pushed into position. Now the great difficulty is to wire the walls together as the soil builds up and yet be sure that the bulldozer, when it pushes in its next load, will not damage or break it.

I fear that I may be accused of overstressing the necessity of using wire as an anchorage whenever possible. What the eye does not see the heart does not grieve over, but I promise you that, if the retaining walls collapse after the effort that goes into making a bank, there would be some truly grieved hearts indeed. The wire not only prevents the walls from being forced outwards but will hold the sleepers in position long after the base of the upright posts have rotted away.

Fill in the Trenches

So now, instead of using 8-gauge wire, use $\frac{1}{4}$ in. plaited cable. First dig trenches about 1 ft. deep and as narrow as possible between the two walls into the soil that the 'dozer has pushed up and along the lines of the straining wire. Fasten one end of each of the wires to the appropriate uprights of one of the retaining walls. Pass the wires along the trenches and through holes that may have to be drilled through the sleepers in the second wall. Just below each hole screw on to the sleeper a gadget called a ratchet wire strainer. These may be purchased from the majority of ironmongers. They consist of a U-bracket in which is mounted a drum with a ratchet attachment. The end of the axle on which the drum and the cog of the ratchet rotate has a squared head which may be wound round with a spanner. The drum is drilled so that a wire may be passed through it. Do this with the cable, take it round the drum and twist it to the in-going length. Pull it as tight as possible before making it fast.

Fill in the trenches and 'doze up the next load of soil and consolidate. Now with the spanner on the squared end of the axle, rotate the drum until the cables are strained tight. By using these ratchet strainers it is possible to re-strain the cables if necessary at a later date. For this reason it is a wise precaution to keep them well greased. The bank is then finished off, the sides shaped and turfed, and the ditches dug and revetted.

The most durable type of all is really a cross between the two I have already described. It may be grass faced on both sides, or sleeper faced on one and grass faced on the other. Which ever, it is a very expensive obstacle to construct. To start, a sleeper box is constructed exactly as described earlier, the only difference being instead of driving half round posts to retain the sleepers, dig in more sleepers to do the job and fill as before. To grass the face, strips of turf must be laid root to herbage. About three feet away from the base of the box start to build a turf wall, being careful to bond the turves in the same manner as laying bricks. As the wall grows, pack the space between it and the sleepers with well-rammed soil, at the same time being careful not to displace the turf.

After the first two feet or so, start to bend the turf wall gradually inwards towards the top sleepers. The bend should become more pronounced as the final intended height approaches. When the turf wall or walls are finished more soil will probably be needed to level off the top. Allow at least 6 months for consolidation before turfing the top in the orthodox way and then it should stand for another 6 months before it is jumped. This is a much sounder way of making a bank but the turf still needs the same attention.

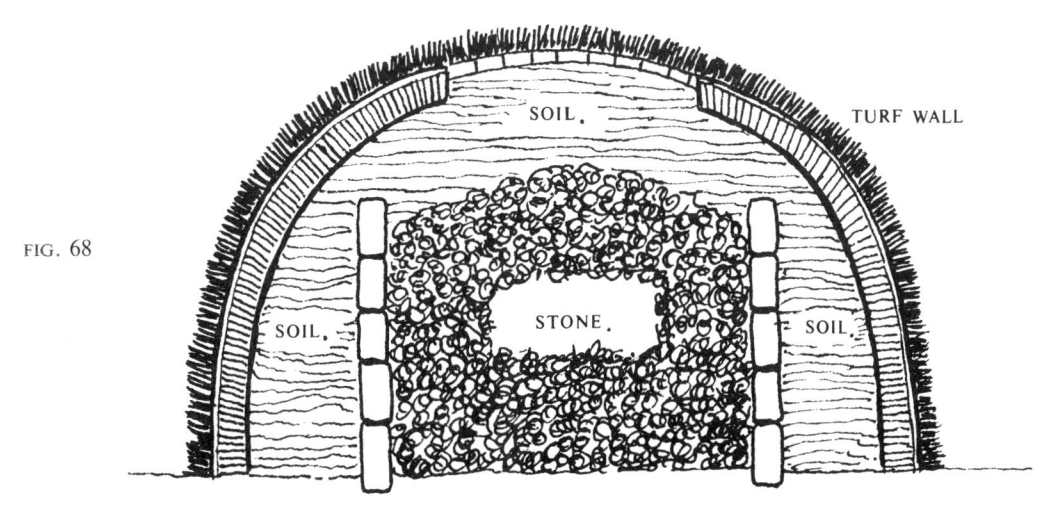

FIG. 68

SOIL. TURF WALL
SOIL. STONE. SOIL.

17   Walls

WALLS, whether they are constructed of stone or brick, are good obstacles to incorporate into a cross-country course. They are a rarity in some parts of the country but common in others. Generally, a certain amount of work is required to strengthen them and to protect the top edges; it is seldom that a wall in its natural state can be jumped without some sort of modification. The only exception is, perhaps, a brick wall with a half-rounded coping, but they are pretty few and far between in the country—in fact, only once in the last twenty years have I been able to wangle one into a cross-country course. The brick wall which was jumped in the European Championships when they were held at Windsor in 1955, was specially constructed for the event, and not many of us can afford to go to that expense!

The majority of walls seen in the country are dry stone walls, that is to say, the stones are fitted together but are not stuck to each other with cement or mortar. The top course of stones is frequently surmounted by bigger, flat stones placed on edge rather like books on a book shelf. These stones, which are called coombers in some parts of the country, invariably have very sharp edges and, of course, are hard and abrasive and would damage a horse severely if he hit them hard.

Top Stones

There are two ways in which I deal with a wall of this type. Firstly, the top stones (coombers) must be removed and then the wall is built up or reduced until it is the required height. Now the top edge on the nearside must be protected, firstly for the horse's sake and secondly to prevent stones from being knocked out if the wall is hard hit. This is done as shown in Fig. 69, by fixing two strong half-round rails to the top of the wall on both sides.

Two stout stakes are driven or dug into the ground on both sides of the wall at the extremities of the part that is to be jumped. Before driving in these stakes, it is as well to explore the ground with a bar, for not infrequently the footings of this type of wall are a bit wider than the wall itself. Thus the stakes cannot be driven so that they are tight against it but will have to be angled so that the end of each stake is touching the top of the wall.

The half-round rails should be positioned between the top of the stakes and the wall so that they just cover the top edge of the top course of stones. In Fig. 69, the half-rails were fixed in a slightly different manner, and this was only because they were protecting a short length of wall rather than just a section of a much longer one. Now drive 6-in. nails through the stakes and into the half-round rails.

Wire Each Pair of Stakes
Between each pair of stakes, and at intervals along the obstacle, form a sort of trench by removing stones from the top two or three courses. Wire each pair of stakes together and the rails at the predetermined intervals with 8-gauge wire, twisting it really tight with a bolt or 6-in. nail and replace the stones. The wall in Fig. 69 was protected by a very heavy telegraph pole and, in order to eliminate the hint of a false ground line, a flower bed was dug at the foot of the wall and planted with hydrangeas provided by Capt. Dick Hawkins, M.F.H., the generous host of the Everdon Horse Trials.

The method of protection which I have just described is inclined to leave a slightly false ground line and, for this reason, I always prefer

FIG. 69

the method depicted in Fig. 70. Unfortunately, it is more expensive, more time-consuming and requires a certain amount of skill. It entails taking off the top five or six courses of stones and then cementing them back into place again. This, of course, is the skilled part of the operation for fitting odd-shaped stones together and finishing up with a flat surface required a good deal of experience or a lot of patience—probably both.

When the wall has been built up to about 4 in. lower than the intended final height, lay two quarter-round (quadrant) rails, one on the near edge and one on the far edge of the top course of stones. On the inside sawn surface of each of these rails, at 18-in. intervals, drive 6-in. nails about 2 in. into the

79

wood and the same distance above their lower sawn surface which is in contact with the top course of stones. Fill the gap between the two quadrant rails with concrete and, if you are a perfectionist, you will surface the last inch or so with cement. (Fig. 70).

Very Important

The object of the 6-in. nails being driven into the quadrant rails is to key the rails to the concrete. If this is not done, the action of rain and frost will eventually free the rails from their moorings.

Occasionally, a cement-bonded stone wall such as may be found round a cattle yard or even a garden (if the owner happens to be away) may be included in a course. All too often, however, it will be higher than the maximum height allowed, but it is perfectly feasible to raise the ground on the nearside by as much as a foot by putting down soil or gravel. It is very important that any building-up of the take-off side should be carried sufficiently far away from the wall to bring it within the height limit when measured from the point of take-off and that it should be given sufficient time to settle before the fence is jumped.

FIG. 70

FIG. 71. *Eridge Horse Trials.*

18 Variation with Logs and Branches

THE more variation in the fences, the more interesting the cross-country, is an axiom which is often easier to appraise than to achieve, for on many park courses there may be very little in the way of natural fence lines to ease the course-designer's task. There are, of course, a number of ways of disguising the straight-forward post and rails to simulate familiar countryside objects, and their method of construction will be discussed later. However, a pile of logs or branches is always a useful stand-by with which to ring the changes and on park courses, at any rate, the material from which to construct them is generally readily available.

Heap the Logs
A log pile may be constructed in two basically different ways. Firstly, and by far the more common way of doing it, is to heap the logs so that they are lying at right angles to the line of the course. If this method is adopted, the size, length and straightness or otherwise of the individual logs is virtually immaterial. The second way is to arrange the logs so that they are lying in the same direction as the line of the course and this means that each one must be fairly straight and of the same length which will conform to the required spread.

The great advantage of the first method is that the fence may be constructed of whatever type of material is available. Figure 71 shows a log pile made from branches which were lying where they had fallen in the nearby wood. Figure 72 shows a fence constructed out of cord wood. Cord wood is the branches of a felled tree which have been cut into short lengths for ease of handling and stacked usually to be sold for cutting up into logs. Each stack is called a cord and, just in case any of you are thinking of going into the business, a cord should contain 128 cubic feet of timber!

Build up the Space
Figure 73 is a picture of a fence, probably made from the trunk of a fallen tree that has been cut up into short lengths and split. In the old days, such lengths would have been a good deal longer but nowadays, when every estate owns a motorised chain saw, it is a good deal less work to cut them into short lengths than trying to lift them if they were longer.

In order to make a fence out of fallen branches (Fig. 71), first decide how wide it is to be and what spread is required at the base; then drive in pegs at the intended position of the four corners. Select two long and fairly straight limbs and position them between the pegs on the near and far side of the fence. It may be that sufficiently long limbs are not available, in which case shorter ones will have to be laid end-to-end. Staple several lengths of 8-gauge wire at intervals along the limb on the far side. The length of each wire should be about 3 ft. greater than the spread of the fence. Staple the other end of each wire to the nearside log directly opposite the original staple, but, in this case, do not drive the staple home.

Now build up the space between these two limbs with more branches, fitting them together as carefully as possible, so that they are stable and leave no hollows, until the intended height is reached. With fencing pliers, strain up the wires attached to the original nearside limb and drive the staples home. More wires are then stapled to it and passed over the completed pile and strained down to the far-side limb.

Using Cord Wood

There must be sufficient of these wires so placed that the surface logs are trapped beneath them and no individual log could be kicked off the fence nor even move. The wires, which must run at right angles to the long axis of the fence, should be stapled to any log that tends to move. If there are any that have not been trapped, they should be fixed to their neighbours with further short lengths of wire. The fence should now be examined to see if there are any holes between the logs in which a horse could possibly catch a leg. If there are, try and fill them with smaller pieces of wood but, if this is impossible, cut turves and ram them in hard.

Using cord wood, much the same procedure is followed but, of course, there will be no pieces sufficiently long to form the near and far foundation. Therefore, this must be done with short pieces which are laid in position to form the required spread. Pegs are then driven well into the ground to prevent these short pieces from moving outwards. Fitting cord wood together is a much more tedious process but, however impossible it looks, it is quite remarkable how, by trying one bit here and another there, it will eventually pack in quite snugly.

A Great Deal of Timber

The split and cut tree trunk fence is also pretty difficult to put together. To make the near and far edges, choose blocks that are as nearly square as possible, building up on them with wedge-shaped pieces and finishing off with rounded ones. The greatest problem with this material is not so much fitting it together, but finishing off at the intended height. The individual blocks are fixed in position with short lengths of wire, stapled so that each block is attached to at least two, if not three, of its neighbours. (Fig. 72.)

FIG. 72. *Everdon Horse Trials.*

FIG. 73. *Windsor Horse Trials, 1968.*

To construct a fence similar to that pictured in Fig. 74 requires a great deal of timber and each piece must be cut to exactly the same length and must be reasonably straight. The only practical way of constructing such a fence is to locate a sawyard that produces pit props for the National Coal Board and persuade them to lend you a load or two. Pit props are generally between 5 and 6 ft. long, so an obstacle made from them will be a pretty big fence. Construction is simple. Two stakes are driven at each extremity to prevent the logs from spreading outwards. They are then built up to within three layers of the intended height. The last three layers should be bedded in sawdust to prevent them from rocking should a horse happen to "bank" the obstacle. The top two layers are fixed by stapling a continuous length of 8-gauge wire first to the cut end of the outside log of the lower layer, then to the outside log of the top layer, then down to the next one in the lower layer and back to the next one in the top layer and so on along the length of the fence. Do this on both the near and far sides.

FIG. 74. *European Horse Trials, Windsor, 1955.*

83

19 Coping with Wire Fences

ALL too frequently, when designing cross-country courses, wire fences are encountered which, somehow or another, must be negotiated. No farmer or landowner likes to have a well-strained wire fence cut and indeed, in these days of high tensile fencing, it is virtually impossible to do so without the whole fence collapsing. Therefore, a jump must be built on the fence line to screen the wire in such a manner that it is impossible for a horse to injure himself or get himself hung up.

Using a gate is one of the quickest and most natural ways of disguising a wire fence. Figure 75 shows a gate serving this purpose at the Burghley Horse Trials several years ago. Firstly, the gate posts are dug in on the far side of the wire fence. The gate is then applied to them on the near side of the fence so that the wires are between the gate and the gate posts. Depending on the height required, the gate may have to be either raised off the ground, or, if it is too high, dug in.

A Word of Warning
Draw out the staples holding the wires from three or four of the fencing stakes on either side of the gate thus reducing the tension and enabling the wires to be positioned behind the bars of the gate. Drill the gate and the posts at the level of the top and second bar

FIG. 75. *Burghley Horse Trials, 1964.*

from the ground. Drive the bolts through the holes but do not tighten until the wires have been positioned.

It will be necessary to add wings and, if posts and rails are used, the posts are driven in on the far side of the wire and the rails then hide the wire and trap it against the posts. A word of warning. It is unwise to erect a gate to be jumped near the gate of the field for, unless the field gate is left open, you can bet your boots someone will jump the wrong one!

Without reinforcement, a gate is not strong enough to be used as an event jump. Apart from the inconvenience, they are expensive and, when they are broken, they are apt to disintegrate altogether. The gate pictured in Fig. 75 was especially made with a strongly reinforced top rail; had it not been, a third post would have been dug in in the centre and cut off flush with the top bar.

Of course, there are many other ways of screening

wire. Any type of solid fence such as a sleeper wall or palisade does the trick. So, too, does a "Tiger Trap" (Fig. 77), the effectiveness of which lies in the fact that the top wire is directly screened by the top rail and the horse is held well away from the lower wires by the diverging uprights.

Wattles are Attractive Fences
A wattle fence (Fig. 78) is impossible to use for screening a wire fence as its proximity would make it extremely difficult to intertwine the individual saplings between the uprights. Wattles are attractive fences, strong and comparatively cheap to make. I have always used either silver birch or chestnut, but I think willow might be quite suitable; apart from these, I doubt if there is any other timber cheap enough and sufficiently pliable to be worth considering. The saplings should be about 25 ft. long with a 3 to 4 in. diameter butt. Do not attempt to make a wattle fence more than 18 ft. wide, for if you do you will find that the tip of the sapling is so thin that it is not worth using or that the butt is so thick that it is impossible to bend.

Dig in four uprights at intervals of 6 ft. and then strain in the first sapling so that the butt is on the near side of one end post. It then passes round the far side of the next post, then the near side and finishes on the far side of the other end post. It may be necessary to use a sledge hammer to get it down to ground level.

FIG. 76

FIG. 77. *Burghley Horse Trials, Pony Club Championships.*

FIG. 78. *Burghley Horse Trials, 1968.*

FIG. 79. *Windsor Horse Trials, 1968.*

FIG. 80. *Windsor Horse Trials, 1969. Tyres used as a drop fence.*

86

The butt of the second sapling will be at the opposite end of the fence to the first and so on until the required height is reached. Nail the top sapling to the top of the posts before cutting the two centre ones flush.

Some of these saplings will take a great deal of bending; indeed, it is at least a two-man job and it is wise for them to work from the same side of the fence so that, if a sapling does slip over the top of a post while being strained into position, it will not do any damage.

A width of 18 ft. is generally quite adequate for the average event fence and wide enough to make wings unnecessary. However, should wings be required, ordinary garden wattle hurdles will add an extra 6 ft. to either end and look thoroughly in keeping (Figs. 78 and 79).

A Shockproof Fence

Fences made out of tyres are quite a novelty and, if neatly made, do not look out of place. But really, their greatest value is for schooling, for it is virtually impossible for a horse to hurt himself by hitting them. Figures 80 and 76 show one way of arranging tyres. These are size 8·25 × 20 which are about the biggest lorry tyres easily obtainable and any tyre agency will be glad to get rid of old ones. About fourteen of them will make a 12 ft. wide fence and, on the flat this will be about 3 ft. 4 in. high, depending upon how much tread is left on the tyres. Thus, to get a maximum-sized fence, they must be sited on an uphill slope or they may be used as a drop fence where maximum height is not required.

These are quite easy fences to make. First dig in two uprights which will form the extremities of the fence. Two rails are laid on the ground at the base of the uprights and they are passed through each tyre as it is put in position. When the space between the two uprights is filled with tyres, one rail is raised as high as it will go and is bolted to the uprights and the second rail is bolted so that it prevents the tyres from rising (see diagram). Although it is impossible to completely eliminate any movement, it will be a shockproof fence which certainly will not come to pieces.

20 Uses for Tractor Tyres and Barrels

TRACTOR tyres are rather more difficult to handle than lorry tyres and, perhaps, a little more difficult to come by, but any dealer in agricultural machinery will be glad to give away badly damaged ones. They are very heavy and are often full of rainwater which is hard to tip out—and makes them even heavier.

Snags Become Obvious
A fence arranged as in Fig. 81 looks a comparatively simple obstacle to make, but as soon as these tyres are stacked on top of each other the snags become obvious. The weight of those on top compresses the ones underneath. The more worn they are, the more they will compress: the less worn they are, the more the top ones will squash the ones beneath so that without supporting them in some way it is impossible to obtain a level top line. Laid on their sides they will produce a spread of up to 4 ft. 6 in. and the finished fence may be of any height, but getting it exactly right is largely a matter of trial and error.

Four Posts
Measure the width of each of the first tyres to be used, which will give an idea of how many will be required and roughly what height each stack will be. If this appears

FIG. 81

to be more than 4 in. greater than the height required, the bottom tyre of each stack will have to be dug into the ground. If less, then adjustments can be made as building proceeds.

Because it is inevitable that the lower tyres will be compressed, the upper wall of each tyre must be supported, which can be done in the following manner. Lay the first tyre on the ground in the required position. Drive four posts into the ground so that they are applied closely to the inside circumference of the tyre and are positioned at right angles to each other. It is important that these posts should be vertical, for the subsequent tyres must be dropped over the top of them and yet still be held firmly in position.

FIG. 82

Bitumastic Paint

Measure the distance between the posts and cut two pieces of wood (say 3 in. by 1½ in.) 1 ft. longer than this distance. Nail one piece to the two near side posts and one to the two far ones so that they are horizontal and supporting the upper wall of the tyre. This process is repeated as each tyre is added, until the last one is put in place. Mark the posts 1½ in. below the level of the underside of the top wall, remove the tyre and saw off the posts at the marks. Then replace the tyre and nail on the cross struts so that they are flush with the top of the posts. Using them as joists, make a "floor" out of 3 in. by 1½ in. thick planks, completely filling the inside circumference of the top tyre so that it is quite impossible for a horse to drop his hind legs into the stack. If it is anticipated that the stack is going to finish up 2 or 3 in. too high, the cross struts supporting the two lower tyres should be nailed on so that they allow the upper walls to be compressed by the required amount. A coat of bitumastic paint finishes off the job and gives the fence a very smart appearance.

Simple but Spectacular

Barrels are usually simple fences to jump but quite spectacular. They have come into prominence recently as a "recognition fence" for the brewers who have been generously sponsoring horse trials in recent years. Figures 82 and 83 show two different methods of using them. In Fig. 82 the barrels are supported on two rails under which beer crates have been packed and there is no possible risk of them being knocked out. However, an arrangement as in Fig. 83 presents a considerable fixing problem, especially if the barrels are on loan and must not be damaged. Those in the photograph were provided especially for the jump and the lower rows were joined together by wire, while a cable was passed through the upper ones and firmly pegged to the ground at either end. If the barrels must be returned intact, a rail pegged to the ground on the near and the far side will hold the lower rows in position, and as long as not more than two-thirds of the diameter of the top row are showing. turf rammed behind them will hold them in position, but if more than two-thirds is showing they

FIG. 83

FIG. 84

must be backed by another rail. I am sure I do not have to stress the point, but it is vital that anything as heavy and perverse as a barrel must be completely secure.

Improving the Appearance

It is unfortunately true that the more a straightforward post and rails is elaborated the more the work and the greater the expense, but nevertheless on a park course where there are few, if any, natural fences, some variation must be introduced. Figure 84 shows a rustic seat which in reality is the conversion of a post and rails with a take-off rail, but it produces something different and is a great deal more interesting. Perhaps one of the most difficult things about making a rustic seat of this size is getting the proportions right—the inclination is to make the bit one sits on too high from the ground.

The safest thing to do is to mock up one end and alter it until it looks right, then take its dimensions and construct the fence to them. To obtain the best effect, fix the top far rail to the top of the posts but bolt the lower rail on to the near side of them, the near side rail being 3 or 4 in. higher than the far one. This gives the correct slope both to the seat and the back rest when the batons are nailed. These should be not less than $1\frac{1}{2}$ in. thick and about 3 in. wide. They must be close enough together to preclude any possibility of a horse getting his foot between them. Leave the near side end posts about a foot higher than the seat and cut a "V" in the top of them to accommodate the arms, also "V" the far end of the arm to fit the far side uprights. Drill them and the near end of each arm and drive in bolts or nails to fix them. Putting on the arms is well worth the extra trouble for they improve the appearance of the fence out of all recognition.

21 Various Forms of Obstacles, including Water

APART from those fences which exploit interestingly uneven ground, it is sometimes extremely difficult to think up any new form of obstacle. However, with a little ingenuity, it is possible to disguise some of the well-tried but mundane fences into something more novel and amusing. The hay rack illustrated in Fig. 85 is really a parallel rails, but, beware, it is rather more difficult to jump than a straight-forward parallel for it has no defined ground line; indeed, it tends to have a slightly false one, but this can be rectified by carefully spreading straw on the ground beneath it.

I have said this before, but I think it is so important to the whole appearance of a course that I will repeat it. If a fence is made to imitate some everyday object, it must really look like that object and be well and neatly made, the only extra cost to achieve this being in time and trouble.

A Natural Island Fence

To construct a hay rack, drive or dig in uprights as if for a parallel rails. Then put in two more, one at each extremity, half-way between the near and far uprights. These two should be much shorter than the others and should finish up a little under one-third the height of the other uprights. The three rows of uprights

FIG. 85

will each carry a single rail which will be notched on to the top of them as previously described.

Fix the middle rail which can be left round, but have the other two rails sawn longways. Lay one half of each in the notches of the near and far uprights and nail the batons from their surfaces down to the centre rail. These batons may be of 3 in. by 1 in. rough timber. They should be close enough together or sufficiently far apart to prevent a horse getting a leg hung up. Apply the other half of each rail to match its pair, so that they have the top ends of the batons held between them. Drill through them and drive bolts into the uprights. Finally,

FIG. 86

height. The tripod effect is obtained by driving a 3 in. by 2 in. stake on both sides of each upright, which is then nailed in position.

If you wish to paint the fence, now is the time to do it, but obviously the bark must be removed from the rail and uprights before this is possible. It is fairly simple to obtain barked timber from most sawyards, but this particular rail was towed along a rough gravel track

FIG. 87

fill the rack with straw. A hay rack is one of the few natural island fences.

Figure 86 is another variation of the parallel rails—a footbridge—which is also a reasonably natural island fence in the right location. When making one of these fences, remember that the bridge must be very near the ground, otherwise the hand rails will be so low that the whole thing will be quite out of proportion.

Figure 87 is a variation of an open ditch, simulating the all too familiar road works. It looks rather complicated but, in truth, it is quite simple to construct. The first thing to do is to put in the inclined uprights in the desired position and bolt the rail to them at the required behind a Land Rover, after which the job was finished by hand.

Now dig the ditch, throwing the soil up below the post and rail, and revett it with sleepers. Most local councils will enter into the spirit of horse trials if tactfully approached and it should be possible to borrow two or three pipes and some red warning lights.

The pipes make an "in-keeping" take-off rail. Pipes can be fixed down in a number of ways, but I think that Fig. 87 illustrates the safest and neatest. The top sleeper of the revetting is fixed so that about half its width is above ground level. The pipes are then applied to it and a rail is passed through them, the two ends of it being strained up to the sleepers with 8-gauge wire. The warning lights add the final embellishment.

There are very few courses that include a natural open water and where the going is durable enough to preclude the need for revetting. Revetting on the near side is perfectly acceptable, but on the far side it can cause serious injury if a horse jumps short and there is any depth of water. Thus, if the far bank must be revetted, the water on the far side must be only a few inches deep and not more than a foot deep to within a yard of it.

Often it is possible to make an open water by damming a stream. It is wise to do this some time before the event to see what result can be obtained and, if this is satisfactory, remove the dam and do not replace it until just prior to the day. In this way the risk of waterlogging the banks is much reduced.

A Jump in Water

When horses are asked to jump into water, the width must be such that it is plain to the horse that he is not expected to jump over it and, of course, the bottom must be level and absolutely sound. The British Horse Society recommend that Novice horses should not be expected to jump into or out of more than 6 in. of water and for Advanced class horses, there should not be more than 20 in. to jump into and 1 ft. to jump out. These figures are issued as a guide and not as a rule. A jump in water should be a post and rails or similar type of fence that the water may pass through. A tree trunk or other solid obstacle is unsuitable as the bow wave set up by the approaching horse builds up against it and is apt to conceal its true dimensions. Fences in water are a good test. They are also helpful to the course designer when a stream has been found with a sound bottom, but in one place one of the banks is sound and the other boggy. A little further down, the reverse may be the case, and it may then be possible to make competitors enter the water from one sound bank, ride down the bed of the stream with a fence so sited that it will hold them on this course until they can get out when the other sound bank is reached.

22 Multiple Obstacles and Alternatives

BEFORE describing the various types of multiple obstacles that may be constructed, I will try and explain the rules which govern their siting and under which they are flagged, numbered and judged. Unfortunately the show-jumping rules are entirely different from the cross-country jump rules but I will briefly mention them so that there can be no confusion by omission. In show-jumping, in the case of double, treble or multiple obstacles, each fence or part of the whole must be jumped separately, on penalty of elimination. When there is a refusal, run-out or fall, the rider must take the whole obstacle again, or he will be eliminated. Doubles, trebles and multiple obstacles carry only one number. If the distance between two obstacles exceeds 12 metres (39 ft. 5 in.), they are regarded as separate fences.

The many variations of multiple obstacles on a cross-country course make it impossible to cover all contingencies by one simple, straightforward rule, and often there is confusion in the correct method of flagging and numbering. On a cross-country course, each fence in a combination is generally considered as a separate obstacle and is flagged, numbered and judged independently. A competitor may refuse twice at each of these obstacles without incurring elimination. He must not, under penalty of elimination, retake any obstacle which he has already jumped.

If, however, an obstacle is formed of several elements, such as banks, steps, ditch and rails or angled obstacles, but designed as one complete test, or with elements so placed that it is impracticable for a competitor to continue at the same point after a refusal, then each element is flagged and is marked with the same number but with an alphabetical suffix; in this case, a competitor who refuses *may* return and retake the complete obstacle.

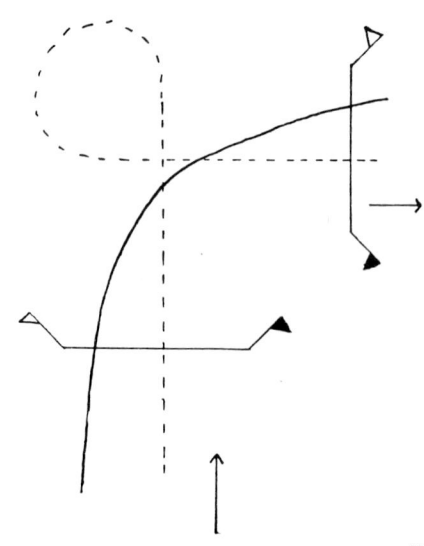

FIG. 88

To illustrate this rule, Fig. 88 shows two fences at right angles to each other. If these two fences are numbered with the same number plus suffix, the track to be followed must be approximately that of the solid line. If, however, the fences are numbered with different numbers,

94

either the solid or dotted line track may be followed without incurring penalty. If the dotted track is to be followed, the horse must not, of course, be presented to the second fence before the circle has been ridden.

Alternative Fences

Figure 89 shows a ditch followed by a post and rails, both being flagged with red and white flags. A horse jumps the ditch and then refuses the rails. If the ditch and the rails are numbered with different numbers, the horse may not recross the ditch between the flags in order to have a second attempt (dotted line). If, however, they are numbered with the same number plus suffix, he may do so (either the solid or dotted line can be followed).

So far, only straightforward single fences have been described. But single fences need not be straightforward, as they may present a varying degree of hazard depending upon the approach. They are called alternative fences and are generally constructed so that the quicker way is the more difficult.

Figures 90 and 91 show two examples. The direction of the course is indicated by the arrows and the alternative routes by the dotted lines. In both cases, the quickest way entails jumping a spread while, to go a slower way, only a single rail has to be negotiated.

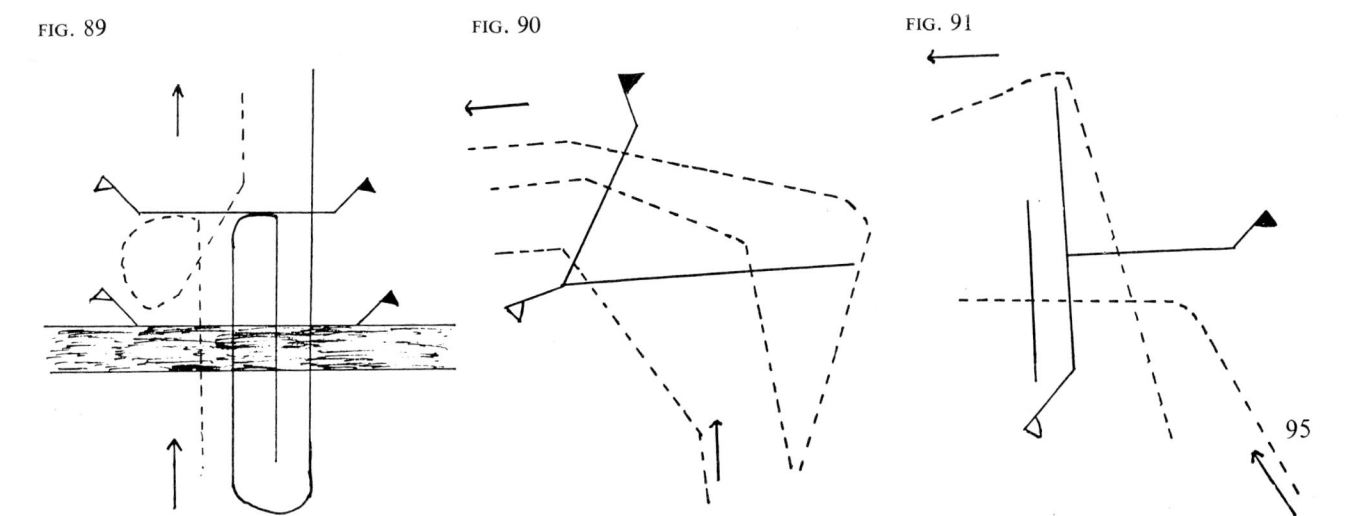

FIG. 89

FIG. 90

FIG. 91

95

From these diagrams it will be seen that a good deal of extra timber is required to construct such fences and yet they add an enormous interest to any course both from the riders' and the spectators' point of view. If, however, all this extra material is to be used, the course builder must try and ensure that it is not going to be wasted. In other words he must be careful not to get the balance wrong and make the difficult part so difficult that no one at all will jump it.

Related Fences

There is no yardstick to help him in this, save experience and "eye". When making these sort of fences, first lay them out on the ground; then, when they appear to be about right, mock them up with light poles until the height and balance of the alternatives seems correct. Then take the exact measurements and construct the fence to them. Do not be in too great a hurry and leave a day or two to fiddle and

FIG. 92

adjust. It is surprising how one's opinion can alter in 24 hours.

All multiple obstacles are known as related fences. Alternative fences **may also be related**. Figure 92 shows the plan of an example of related alternative fences. In this obstacle, the balance between the various routes is obtained by the distances between the elements as well as by height. The distance between "A" and "B" would be awkward—say 18 ft. That between "C" and "D" would be easy, 25 to

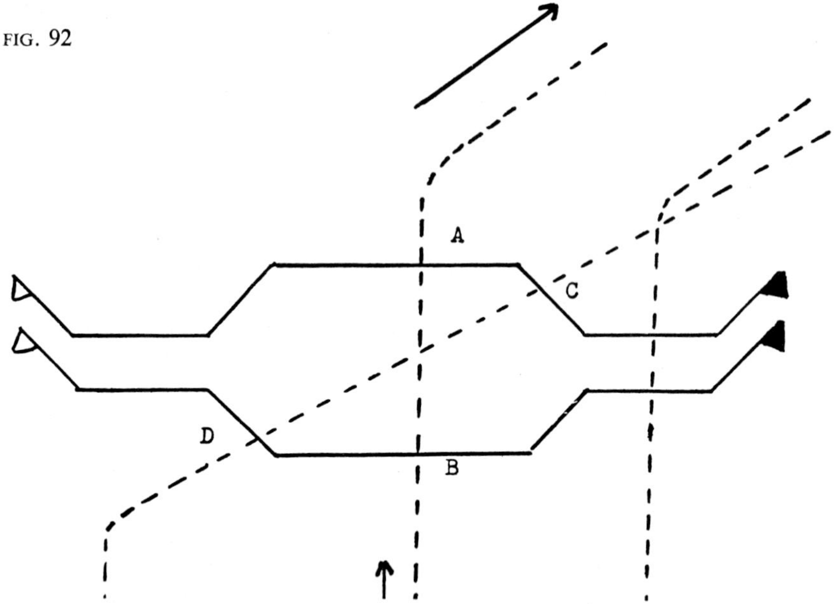

27 ft., while the parallel rails at each end would have a spread of, say, 5 ft. You pays your money and you takes your choice!

The concept of these fences may be carried even further to what might be called dependent alternative fences. These are generally multiple obstacles and the route taken over the first of them will dictate which of the subsequent obstacles will be jumped.

Two dependent alternative fence patterns are illustrated in Figs. 93 and 94. The line "A—B" in Fig. 93 is a high and unjumpable thorn hedge so that, once the first of these three obstacles has been jumped, the rider is committed to either the left- or right-hand route.

Difficult to Balance

In Fig. 94, which really only comprises two fences, much the same applies. "A—B" is an unjumpable post and rail running up to a tree. If the rider jumps the first fence to the left of "B" he is bound to take on the parallel rails. If, on the other hand, he jumps to the right, he has to go through the parallel rails and jump over the stile at the far end of them—a much slower but easier process.

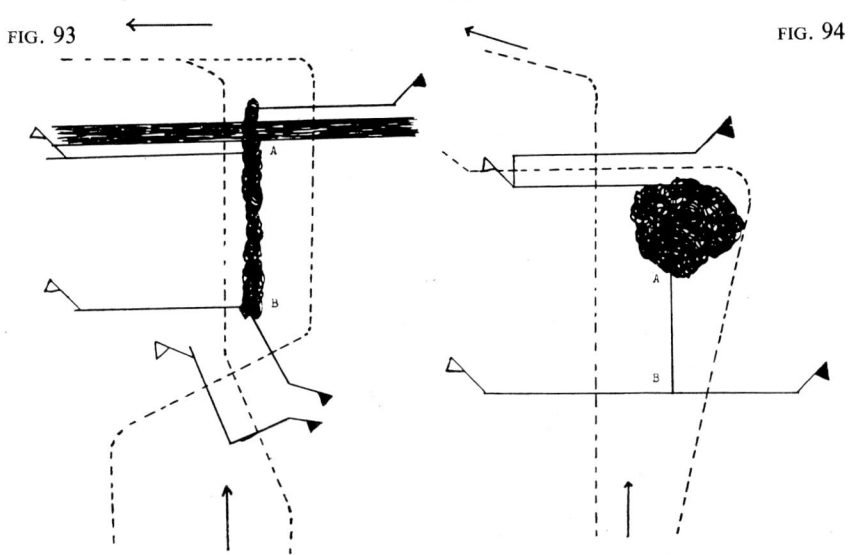

Notice the position of the flags on the parallel rails. They are so placed that, whichever route is taken, the white flag is on the left and the red on the right—as indeed they must be. Also there must be some way of quickly dismantling the stile for, if a horse refused it, some difficulty might be experienced in getting him to back away down the parallel rails.

These are the most difficult fences of all to balance correctly and the most wasteful if they are balanced incorrectly. If, in the jumping of any alternative fence, 25 per cent of the competitors go one way and 75 per cent the other, you may be sure the course constructor has done a first-class job.

FIG. 93

FIG. 94

97

23   Knots, Ropes, Bolts and Other Equipment

EVERYONE, I should think, knows how to tie a reef knot, but not everyone knows why it is a better knot, of its kind, than any other. The answer is, of course, that however hard it is strained, it will never slip nor will it ever bind, which means it will always be easy to untie. This is a basis upon which all recognised knots are designed and it is worth learning how to tie one or two of them in order to save the time, temper and finger nails that are wasted on trying to undo a botched up muddle of solid hemp.

I find that I seldom use any other knot than a bowline (Fig. 95) or a clove hitch (Fig. 96). The former, which is very similar to a reef knot, forms a non-slip loop, which can quickly be converted to a slip knot by merely doubling end "A" through it. The clove hitch is used mainly for anchoring a rope to a post or a tree, although it does have other uses.

FIG. 95

It has the merit of being one of the simplest knots in the book and, when tension is taken off it, it virtually undoes itself.

Roping an Arena

Figure 97 is the method I use

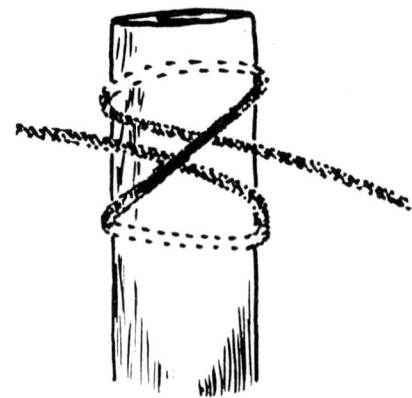

FIG. 96

when roping a cross-country course or dressage or jumping arena. If rope is left up for any length of time it is very apt to "hang out" and appear to stretch. If tied to every post it is a very tedious business taking up the slack. With the knot illustrated, when "A" is eased back towards "B", it is perfectly simple to tighten the rope beyond it, which is then locked in position when the rope beyond "A" is tightened. With a bit of practice, it is comparatively easy to put rope round a post in this manner, using

98

FIG. 97

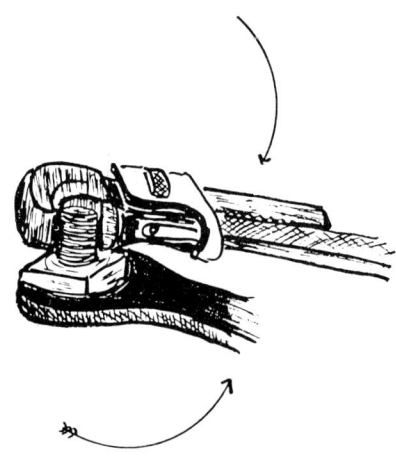

FIG. 98

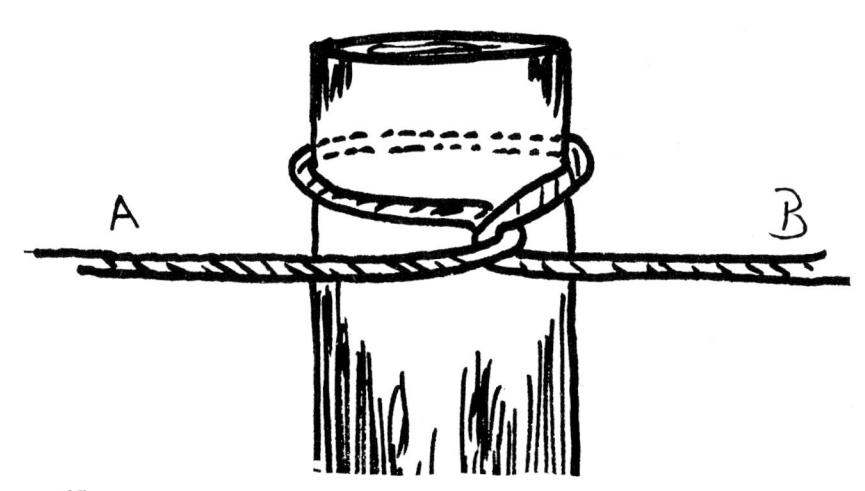

only one hand which leaves the other free to carry the remainder of the coil.

This booklet has been devoted to the construction of fences, but I think it will not come amiss to mention a few of the snags that arise when dismantling unwanted fences but, at the same time, preserving the timber from which they were made.

Getting the Top Rail Off
The first job is to get the top rail off, but nine times out of ten it will be found that the nuts have rusted to such an extent that the collar of the bolt has insufficient grip to hold it and it will rotate with the nut when a spanner is applied. Depending on the circumstances, there are two ways of coping with the problem, but whichever is used the first step is to squirt the nut with penetrating oil and leave it overnight. I have found that the most convenient brand to use is "Plus Gas". If there is over a quarter of an inch of bolt protruding through the nut, it is possible to fix a pair of stilsons to it and hold it firmly while the nut is undone with a spanner until it winds the stilsons off the end of the bolt (Fig. 98). Now

99

FIG. 99. *Pulling out the uprights.*

the bolt can probably be knocked back so that the stilsons may be applied to the head and the nut finally removed. Most fence constructors, however, cut off any surplus bolt, leaving it flush with the nut so there is no protrusion on which a horse might injure himself. In this case, there is no alternative but to cut the nut in half by hack-sawing down the long axis of the bolt (Fig. 99) and then knocking the two halves off with a hammer.

The quickest way to pull out the uprights is with a hydraulic lift on a tractor but, if no tractor is available, bolt together two 9 ft. long rails about a foot from one end. These rails should be about 4 in. in diameter. Fix a rope low down on the post to be removed (with a clove hitch!). Spread the feet of the two rails which are then leant over the top of the post, pass the rope through the top "V" and fasten it to a vehicle. When traction is applied, the direction of the pull will be converted from near horizontal to near vertical as the crossed rails move into an upright position. The post should then come out with the minimum disturbance to the surrounding ground (Fig. 100).

Various Tools

It is probably better to fix the rope to the front of the vehicle and reverse away so that it can be stopped immediately if anything goes wrong. If a Land Rover is used, do not tie the rope to the bumper without protecting the edges as they are sharp enough to cut through quite thick rope.

Before finishing, I will mention the various tools that I think are necessary for constructing courses. I will list the bare minimum and then catalogue those that are not essential but make life easier. However, before starting, let me emphasise that when buying tools, it always pays, in the long run, to buy the best, so

FIG. 100

long as they are going to be properly cared for, and kept sharp and rust free. The essential heavy tools are:

Spade, the cutting edge of which should be periodically sharpened by rubbing up with a stone. Bar, about 10 lb. in weight and about 4 ft. 6 in. long with a chisel point at one end and a pencil point at the other. 7-lb. sledge hammer, 14-lb. post hammer. Heavy-duty, long-handled axe.

A Claw to Your Taste

Now for the essential light tools: Bow-saw, the blade of which should not be less than 3 ft. long so that it may be used by two people without fear of trapping each other's fingers. Hacksaw with an adjustable frame to take any length of blade. Hacksaw blades, and as they snap easily it is always worth carrying at least half a dozen. Regrettably, there is no doubt in my mind that those made in Sweden are the best and that applies to all saws. Hammer—these are made in different weights and with different lengths of shafts. Choose one with a claw to your own taste. I have found that Stanley's steel-shafted are the most durable.

When Drilling

Carpenter's brace, which is sold in a wide range of models and an equally wide range of prices! I would recommend one with a ratchet and not less than a 10 in. sweep, preferably 12 in. When drilling, a ratchet enables the operator to keep the bit rotating in the same direction when making a series of half-turns on the crank. Not infrequently, a situation arises when it is impossible to make a full revolution of the crank because it is obstructed by some other part of the fence. A 12 in. sweep indicates that the crank is set back 6 in. and when in use the operator's hand will describe a circle 12 in. in diameter. The greater the circle, the more the power and the less the wear and tear on the biceps! **9/16th in. wagon bit, the longer the better, but should be between 10 and 12 in.**

For Drawing Nails

A spanner with jaws 1 in. wide, which fits on to the nut of a $\frac{1}{2}$ in. coach bolt. A garden hook for cutting undergrowth. Long-handled secateurs for trimming natural hedges or artificial birch ones. A stone and a small file for sharpening tools and bits. A 12 ft. spring measure and a 25 ft. tape measure. Fencing pliers, which have a variety of uses such as straining and cutting wire and drawing staples. A wrecking bar, often called a jemmy, not less than 2 ft. long, with a chisel point at one end and swan-necked claws on the other. This tool is quite invaluable for drawing nails or bolts and dismantling fences both cross-country and natural. A spirit level.

Supplementary Heavy Tools

I also carry fence props, coach bolts from 5 in. to 16 in. in length, 6 in., 5 in. and 4 in. nails. $1\frac{1}{2}$ in.

staples, $\frac{3}{4}$ in. felt nails which are useful for putting up hardboard notices or nailing fence flags on to posts, and 8-gauge wire, which is sold by weight—56 lb. is quite a lot of wire.

The following are supplementary heavy tools which under certain circumstances can speed up course-construction and reduce manual work. Pickaxe, which does nothing that cannot be done with a bar, but some jobs it does rather more quickly. A scythe or long-handled hook takes the back-break out of clearing large areas of undergrowth. A slasher is a long-handled bill-hook with which it is possible to cut branches as high as 10 or 12 ft. off the ground.

A drive-all, which is a post-driver designed for two people (or more) to use. It consists of about 3 ft. of wide-diameter pipe, closed at one end and provided with handles on the sides. The post is inserted into the pipe and driven into the ground by the operators raising and lowering the pipe so that the closed end strikes the top of the post. The great merit of this tool is that it never damages the posts so that they may be used time and time again for jumping arenas and the like.

The Value of Timber Chain

Timber chain, which is a 5 ft. length of chain fitted with a hook on one end and a ring on the other. The chain should be made of not smaller than 2 in. links. It has a hundred and one uses—from towing out bogged cars to straining posts back to rails before bolting them. Hand earth auger which really is a luxury, for it only works properly in stone-free soil. There are a number of designs but they all work on the same principle. They are wound into the ground for 6 or 7 in., then when pulled up bring the soil out with them. They will dig a hole up to 9 in. in diameter and 3 ft. deep.

Supplementary hand tools might include a 1 in. cold chisel, which is so hard that it may be hammered through 6 in. nails and small bolts. A Surform file which is a wonderful tool for rounding off sharp edges and removing protruding knots. 12 in. stilsons, one of their uses having been discussed earlier in this article. They will firmly grip anything made out of any material up to $2\frac{1}{2}$ in. wide.

A hand axe and a pair of garden shears are also useful adjuncts, and it is surprising how often a small portable vice can save a lot of time and trouble. A steel wedge about 2 in. wide and 6 in. long, used in conjunction with a 7-lb. sledge hammer, will take almost anything apart. Earlier I suggested the purchase of a 25 ft. tape measure, but one 100 ft. long will save an immense amount of time. They are much more expensive but they are more accurate when laying out multiple fences and save a lot of walking when setting up a dressage arena.

There are three mechanical aids that I have not yet mentioned, for few people could afford to buy them unless they also have some other use for them. The first is a petrol-driven chain saw—they are very expensive, but save a tremendous amount of work, and when making fences out of really heavy timber they can be very accurate. They are dangerous; always treat them with respect.

To use either of the other two it is necessary to have a tractor. I think that the better of them is the Tompkins Postmaster, which will drive sleepers or posts up to 7 in. in diameter as far into the ground as may be desired. It is a tool that must be treated with great caution. The alternative is the power-driven earth auger which, like the Postmaster, fits behind a tractor; but this tool actually digs out a hole up to 3 ft. deep and, within reason, to whatever diameter is desired, according to the size of auger used. The use of these tools will cut the time it takes to make a course by 75 per cent and, if you are working on rocky ground, they save the frustration that spoils the fun of making a course.